Planck's Columbia Lectures

Delivered at Columbia University
in 1908

First Translated and Published in 1915

Unabridged and Abridged Versions

With Commentary

by

Dr. Weldon Vlasak

Planck's Columbia Lectures

Abridged and Unabridged Versions

With Commentary by Dr. Weldon Vlasak

Published in 2005 by:

Adaptive Enterprises
Clatonia, NE 68328

First Edition 2005
First Printing 2005

ISBN Number:

0-9659176-3-0

Dedicated to my vivacious,
sharp, intelligent, quick-witted,
kind, caring, heartfelt, thoughtful,
and most wonderful wife

Marge

- who put up with me
all of these years -

Acknowledgements

To Professor Max Planck

to whom the world owes a great debt

(The graphs in this book were plotted

using the Mathcad computer program)

Planck's Columbia Lectures

Table of Contents

Planck's Columbia Lectures

Preface

I hadn't originally planned to write this book, being an engineer, not a physicist. It all began in 1983 when I went to a flea market and purchased a number of scientific texts on physics and chemistry. The physics of atoms and the mystery of the gravitational force had intrigued me since my first course in chemistry, and this was a chance to study the physics of gravity from a real world perspective. The material on atomic physics in these books was based on the characteristics of hydrogen, which is the simplest atom. Most of the universe is made up of hydrogen (about 80%), so if one can understand hydrogen, presumably one will have a pretty good start on the physics of the universe.

The hydrogen atom has but a single proton and a single electron, and the electron is moving in a path around the proton. The thought came to me: "Why not analyze it as an electronic circuit?" Other methods had been used in the texts that I had purchased, and there were some discrepancies in the mathematical analysis, where the equations on one page couldn't be substituted into the equations on the next page without creating contradictions. It is a long story, but my studies eventually led to some new ideas and stimulated me to writing three books on the subject (*The Secret of Gravity*, *The Electric Atom* and *Secrets of the Atom*), and some technical articles for scientific publications.

Quantum theory was of little interest to me during this time, since this method is based primarily on energy concepts, and I was using a quite different analytical approach based on forces. I became involved in an Internet discussion with a group of postgraduate physics students in Sydney, Australia, about the new physical concepts in my books. In order to draw my own conclusions about physical concepts, it was necessary to challenge the assumptions upon which contemporary theories are based. It seemed to me that theories based on force concepts could lead to greater perceptions than those based on energy. However, these bright young students had insisted that deriving equations that are based on energy concepts was the better approach. They were so sincere and insistent in their beliefs that I eventually decided to search the archives for a complete description of the original quantum theory, which was, luckily enough, found at a local university library. The quality of the material in Planck's recorded lectures is excellent and quite fascinating, and a great deal of time was spent in evaluating the differing perspectives of the mathematical methods against those of real world physics.

Although Planck had based his theory on the laws of conservation of *energy*, while my analysis is based on electric *force* fields, no significant conflicts between the two theories were discovered. Surprisingly, Planck had adopted a similar (force-based) approach in another theory of his that seems to have received little attention. This theory is described in detail in one of these lectures.

My studies of Planck's theory led to a series of articles that I wrote for "Chemical Innovation" (a publication of the American Chemical Society) and a technical presentation of a paper about Planck's theory at a scientific symposium. Sometimes, some rather strange questions get asked. At the end of the presentation, a university professor asked me "...how did he get the [radiation] curve to bend over?" Actually, Wien was the one who had first accomplished this feat, but Planck derived the solution method. It took a moment to figure out what he was talking about, and my response was: "Oh, you are talking about Stirling's equation". He just looked to me, scoffed and walked away. Stirling's equation used in one of these lectures, and it is an important part of Planck's derivation of his radiation theory.

In my college physics text, the author offered his opinion that Planck's radiation equation was "...still empirical at that stage and did not constitute a theory". Planck's modification does appear to be of a minor nature that could very well have been the result of empirical modeling. However, after studying Planck's Columbia lectures, I have reached quite a different conclusion and a great admiration for the abilities of Professor Planck.

Chapter 9 contains a summary of all of the lectures. There is some repetition with respect to the commentaries at the end of each lecture, but some readers may prefer to look there for a shorter and easier read. Note that the last two of his lectures include material that extends beyond his radiation theory.

In Chapter 10, it is shown how Planck's analysis fits together with important modern scientific theories. It also includes a few of my own theories, and the correlations with Planck's radiation theory are discussed. There is a new description of the transverse nature of radiation that has some surprises[1], and the characteristics of thermal noise is defined in a way that is quite different than you will find in other texts.

The original equations in the lectures were not all numbered, and this method has been preserved, along with most everything else in presenting the lecture information in its original form. In order to maintain consistency, this numbering method is maintained throughout the first nine

1 "A Different Picture of Radiation", IEEE Antennas and Propagation Society International Symposium 2003, Columbus, Ohio.

chapters. There are not a large number of equations, as compared to contemporary physics texts, and most capable electronic engineers should have sufficient scientific background to be quite familiar with the mathematical methods.

There is more knowledge and insight about the real world of physics in Planck's recorded lectures than any other technical paper or textbook that I have read in my long career. This is a book that can be studied over and over in order to gain further knowledge of how atoms and molecules of matter, in various forms, react and the effects that occur. Even a small amount of the information that can be gleaned from this book is quite worthwhile. If you are a science student and devote sufficient effort in studying Planck's lectures, it can lead to a greater perception of physical phenomena. This might also the case for university professors and research scientists.

What follows is a class study in scientific methodology.

Planck's Columbia Lectures

Introduction

You are about to read about one of the most thorough scientific analyses that were ever conducted, as described by one of the greatest scientists in history. It is the original version of quantum theory that has sometimes been misunderstood and inaccurately represented. Planck's scientific approach was based on measurements, physical laws, logic, inductive reasoning, reality, mathematical analysis, and clear thinking. There is much to be learned from this historic document, which can be studied over and over, and a greater understanding of physics will be gleaned each time, since there is much to be learned. The analytic methods used in these lectures will prove be useful for improving the analytical skills of any engineer, scientist, mathematician or student.

Max Planck first presented his theory to the Berlin Physical Society in 1900. Planck delivered a series of lectures at Columbia University in 1908 in which he described his famous quantum theory. It is fortunate that these lectures were transcribed, translated (from German to English) and subsequently published in 1915. I believe that it is important to make this information available to the general public.

The presently accepted methods of *quantum mechanics*, developed over the past century, deviate markedly from Planck's original quantum theory. Many recognized scientific experts in this field admit that they do not fully comprehend today's *quantum mechanics* theory. In contrast, the detailed descriptions of Planck are brilliantly clear, are quite detailed, and each term of each equation is defined. Also, there is not the proliferation of equations that is found in modern texts on quantum mechanics. Planck goes to great lengths in simplifying his equations for greater comprehension, which is a refreshing difference from the content of physics texts that usually have a very large number of equations that are not always well defined.

Planck has been given credit for the “-1” in the denominator of his radiation equation; a seemingly minor accomplishment. How did he formulate his radiation model? How did the “-1” term end up in the denominator of his equation? Is Planck’s radiation equation a simple empirical model? How did he get the curve to bend over? These questions are answered in this book, and the answers are surprising and impressive. You will also discover a remarkable set of scientific methodologies that will be of great use to any scientist.

The differences between Wien’s earlier radiation model and that of Planck do not appear to be very significant, as is seen in the graph below.

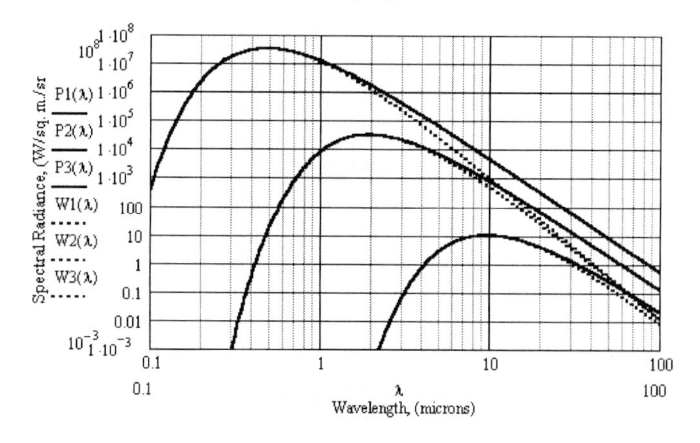

Figure 0.1 Plots of Planck’s Blackbody Radiation Equation

In the above graph, Planck’s radiation curves are the solid

lines, while Wien's curves are dotted. The differences that occur at the longer wavelengths is due to the "-1" term in the denominator of Planck's radiation equation,

$$\Re_\lambda = \frac{c_1}{\lambda^5} \frac{1}{e^{c_2/\lambda T} - 1} \; .$$

Planck's constant, *h*, which appears in many equations in physics, is contained within the constant c_2.

In my college physics text, the author stated his belief that Planck's radiation equation was "...still empirical at that stage and did not constitute a theory". This seemed reasonable at the time, since the modification does appear to be of a minor nature and could very well have been the result of empirical modeling. After studying Planck's original Columbia lectures, I have reached quite a much different conclusion. Planck devised a theoretical and philosophical theory that fits the reality of the universe in all respects.

The concept of "entropy" is basic to the development of Planck's theory. This term discussed in most physics and chemistry texts. How many scientists fully understand this concept? One scientist recently claimed that "...the entropy of an entire system no longer equals the sum of the entropies of the individual parts" [for reactions not in equilibrium]. See if you agree with this assertion after reading the first three chapters.

The first eight chapters of this book cover each of his actual translated lectures. The short commentary at the beginning of each chapter mentions certain points that might be overlooked, primarily because of the long sentences in some paragraphs which are typical of early German scientific syntax. Some amount of repetitive readings of some sentences may be required to grasp the true meaning, although none of this material is to be regarded as superfluous. Planck has done a remarkable job of explaining a highly complex theory in fine detail and with great simplicity, clarity and high efficiency.

The majority of the comments are placed in the summary at the end of each chapter so that readers do not have to be influenced by my observations, interpretations and conclusions, if they choose not to do so. However, the commentaries are basically complete and may provide greater perspective for those who may experience difficulty with the style of the translation. Therefore there is repetition of the unabridged version that is present in the abridged version of Chapter 9.

In contemporary scientific writings, the word ***quantum*** is thrown around a lot, like pickles being tossed into a barrel. Evidently, the more "quantums" that are included in a technical article or presentation, whether they are needed or not, the more likely it is to be published. Planck did not use the work hardly at all, although his definition of state space leads is a basic quantum concept. Some readers may attain a new respect for the word *quantum* after studying the material that follows, and it will then have much more meaning.

The last two of Planck's lectures extend beyond his radiation theory. In the seventh lecture, he presents another very important theory that seems not to have gotten much attention. The eighth lecture is a discussion of Einstein's theory of relativity, and, although not extensive, it is clear, concise and, I believe, accurate.

Chapter 9 contains a summary of all of the lectures. It is therefore, in itself an abridged version of Planck's theory, with some additional analysis and commentary. However, there is not much missing from the unabridged version, and it will most likely be easier to read, due to the difficulties in the translation.

In Chapter 10, it is shown how Planck's quantum theory fits together with other important scientific theories. It also includes a summary of some of my own scientific investigations. Oliver Heaviside's contributions, which are not to be underestimated, also fit together with Planck's theory, leading to further definition of the impulse function. This then leads to the connection between Planck's radiation theory and electrical noise, and a new definition of noise is revealed.

There is also a description of the transverse nature of radiation from a technical paper that I presented at the *IEEE Antennas and Propagation Society International Symposium 2003*. These various phenomena are woven together in a set of theories that are compatible in Chapter 10. The assumptions upon which commonly accepted theories are challenged, and the resulting unique and different conclusions that are reached were unexpected.

The lectures are formatted in a form that is as close as possible to that of the original printing, including the equations and drawings, which are somewhat different from the common practices used today in technical writing. For instance, Planck did not number all of his equations, even though they are not large in number. In order not to cause confusion, this practice is maintained in the commentaries at the end of each chapter and in Chapter 9.

Planck's recorded lectures contain more knowledge and insight about the real world of physics than any other technical paper or textbook that I have ever before encountered. This is a book that can be studied over and over in order to glean additional knowledge of how the atoms and molecules of matter of various forms react and the effects that occur. Even a small fraction of the information that is available in Planck's lectures is worthwhile, and, if you are a science student, you could end up knowing more about Planck's original theory than your college instructor. In fact, this book would be an excellent choice for the text of a one or two semester college course for physicists, chemists and engineers. What follows is a remarkable case study in scientific methodology.

Planck's Columbia Lectures

Chapter 1

Introduction to the First Lecture

Planck begins this first lecture by establishing the foundation of his theory, which is based on real world sense perceptions, physical definitions of all parameters and philosophical conclusions. He emphasizes the importance of the principles of the conservation of energy and the laws of thermodynamics in relationship to the "energy principle" and discusses the energy relationships associated with the concept of perpetual motion.

The paragraph in which he discusses the "unification of our system of physics" is quite interesting and is discussed further in later lectures. There are a few important things to keep in mind. His description of the principles of "reversibility" and "irreversibility" are used throughout most of his lectures. Much of what follows is based on what he defines as the quantity called "entropy" and how it relates to how these two fundamental processes and the state of a physical system. There are only five simple arithmetic equations in this chapter that are used to begin to establish definitions of entropy and irreversibility. The basic relationships between entropy and reversible and irreversible processes are fundamentally established by the "inequality of Clausius", which is a function of only two variables.
However, Planck provides another, more substantial definition of entropy in subsequent lectures.

Planck's Columbia Lectures

FIRST LECTURE

INTRODUCTION
REVERSIBILITY AND IRREVERSIBILITY

Colleagues, ladies and gentlemen: The cordial invitation, which the President of Columbia University extended to me to deliver at this prominent center of American science some lectures in the domain of theoretical physics, has inspired in me a sense of the high honor and distinction thus conferred upon me and, in no less degree, a consciousness of the special obligations which, through its acceptance, would be imposed upon me. If I am to count upon meeting in some measure your just expectations, I can succeed only through directing your attention to the branches of my science with which I myself have been specially- and deeply concerned, thus exposing myself to the danger that my report in certain respects shall thereby have somewhat too subjective a coloring. From those points of view which appear to me the most striking, it is my desire to depict for you in these lectures the present status of the system of theoretical physics. I do not say: the present status of theoretical physics; for to cover this far broader subject, even approximately, the number of lecture hours at my disposal would by no means suffice. Time limitations forbid the extensive consideration of the details of this great field of learning; but it will be quite possible to develop for you, in bold outline, a representation of the system as a whole, that is, to give a sketch of the fundamental laws which rule in the physics of today, of the most important hypotheses employed, and of the great ideas which have recently forced themselves into the subject. I will often gladly endeavor to go into details, but not in the sense of a thorough treatment of the subject, and only with the object of making the general laws more clear, thorough appropriate specially chosen examples. I shall select these examples from the most varied branches of physics.

If we wish to obtain a correct understanding of the achievements of theoretical physics, we must guard in equal measure against the mistake of overestimating these achievements, and on the other hand, against the corresponding mistake of underestimating them. That the second mistake is actually often made, is shown by the circumstance that quite recently voices have been loudly raised maintaining the bankruptcy and, debacle of the whole of natural science. But I think such assertions may easily be refuted by reference to the simple fact that with each decade the number and the significance of the means increase, whereby mankind learns directly through the aid of theoretical physics to make nature useful for its own purposes. The technology of today would be impossible without the aid of theoretical physics. The development of the whole of electro-technics from galvanoplasty to wireless telegraphy is a striking proof of this, not to mention aerial navigation. On the other hand, the mistake of overestimating the achievements of theoretical physics appears to me to be much more dangerous, and this danger is particularly threatened by those who have penetrated comparatively little into the heart of the subject. They maintain that some time, through a proper improvement of our science, it will be possible, not only to represent completely through physical formulae the inner constitution of the atoms, but also the laws of mental life. I think that there is nothing in the world entitling us to the one or the other of these expectations. On the other hand, I believe that there is much which directly opposes them. Let us endeavor then to follow the middle course and not to deviate appreciably toward the one side or the other.

When we seek for a solid immovable foundation

which is able to carry the whole structure of theoretical physics, we meet with the questions: What lies at the bottom of physics? What is the material with which it operates? Fortunately, there is a complete answer to this question. The material with which theoretical physics operates is measurements, and mathematics is the chief tool with which this material is worked. All physical ideas depend upon measurements, more or less exactly carried out, and each physical definition, each physical law, possesses a more definite significance the nearer it can be brought into accord with the results of measurements. Now measurements are made with the aid of the senses; before all with that of sight, with hearing and with feeling. Thus far, one can say that the origin and the foundation of all physical research are seated in our sense perceptions. Through sense perceptions only do we experience anything of nature; they are the highest court of appeal in questions under dispute. This view is completely confirmed by a glance at the historical development of physical science. Physics grows upon the ground of sensations. The first physical ideas derived were from the individual perceptions of man, and, accordingly, physics was subdivided into: physics of the eye (optics), physics of the ear (acoustics), and physics of heat sensation (theory of heat). It may well be said that so far as there was a domain of sense, so far extended originally the domain of physics. Therefore it appears that in the beginning the division of physics was based upon the peculiarities of man. It possessed, in short, an anthropomorphic character. This appears also, in that physical research, when not occupied with special sense perceptions, is concerned with practical life, and particularly with the practical needs of men. Thus, the art of geodesy led to geometry, the study of machinery to mechanics, and the conclusion lies near that physics in the last analysis had only to do with the sense perceptions and needs of mankind.

In accordance with this view, the sense perceptions are the essential elements of the world; to construct an object as opposed to sense perceptions is more or less an arbitrary matter of will. In fact, when I speak of a tree, I really mean only a complex of sense perceptions: I can see it, I can hear the rustling of its branches, I can smell its fragrance, I experience pain if I knock my head against it, but disregarding all of these sensations, there remains nothing to be made the object of a measurement, wherewith, therefore, natural science can occupy itself. This is certainly true. In accordance with this view, the problem of physics consists only in the relating of sense perceptions, in accordance with experience, to fixed laws; or, as one may express it, in the greatest possible economic accommodation of our ideas to our sensations, an operation which we undertake solely because it is of use to us in the general battle of existence.

All this appears extraordinarily simple and clear and, in accordance with it, the fact may readily be explained that this positivist view is quite widely spread in scientific circles today. It permits, so far as it is limited to the standpoint here depicted (not always clone consistently by the exponents of positivism), no hypothesis, no metaphysics; all is clear and plain. I will go still further; this conception never leads to an actual contradiction. I may even say, it can lead to no contradiction. But, ladies and gentlemen, this view has never contributed to any advance in physics. If physics is to advance, in a certain sense its problem must be stated ın quite the inverse way, on account of the fact that this conception is inadequate and at bottom possesses only a formal meaning.

The proof of the correctness of this assertion is to be found directly from a consideration of the process of development which theoretical physics has actually undergone, and which one certainly cannot fail to designate as essential. Let us compare the system of physics of today with the earlier and more primitive system which I have depicted above. At the first glance we encounter the most striking difference of all, that in the present system, as well in the division of the various physical domains as in all physical definitions, the historical element plays a much smaller role than in the earlier system. While originally, as I have shown above, the fundamental ideas of physics were taken from the specific sense perceptions of man, the latter are today in large measure excluded from physical acoustics, optics, and the theory of heat. The physical definitions of tone, color, and of temperature are today in no wise derived from perception through the corresponding senses; but tone and color are defined through a vibration number or wave length, and the temperature through the volume change of a thermometric substance, or through a temperature scale based on the second law of thermodynamics; but heat sensation is in no wise mentioned in connection with the temperature. With the idea of force it has not been otherwise. Without doubt, the word force originally meant bodily force, corresponding to the circumstance that the oldest tools, the ax, hammer, and mallet, were swung by man's hands, and that the first machines, the lever, roller, and screw, were operated by men or animals. This shows that the idea of force was originally derived from the sense of force, or muscular sense, and was, therefore, a specific sense perception. Consequently, I regard it today as quite essential in a lecture on mechanics to refer, at any rate in the introduction, to the original meaning of the force idea. But in the modern exact definition of force the specific notion of sense perception is eliminated, as in the case of color sense, and we may say, quite in general, that in modern theoretical physics the specific sense perceptions play a much smaller role in all physical definitions than formerly. In fact, the crowding into the background of the specific sense elements goes so far that the branches of physics which were originally completely and uniquely characterized by an arrangement in accordance with definite sense perceptions have fallen apart, in consequence of the loosening of the bonds between different and widely separated branches, on account of the general advance towards simplification and coordination. The best example of this is furnished by the theory of heat. Earlier, heat formed a separate and unified domain of physics, characterized through the perceptions of heat sensation. Today one finds in well nigh all physics textbooks dealing with heat a whole domain, that of radiant heat, separated and treated under optics. The significance of heat perception no longer suffices to bring together the heterogeneous parts.

In short, we may say that the characteristic feature of the entire previous development of theoretical physics is a definite elimination from all physical ideas of the anthropomorphic elements, particularly those of specific sense perceptions. On the other hand, as we have seen above, if one reflects that the perceptions form the point of departure in all physical research, and that it is impossible to contemplate their absolute exclusion, because we cannot close the source of all our knowledge, then this conscious departure from the original conceptions must always appear astonishing or even paradoxical.

There is scarcely a fact in the history of physics which today stands out so clearly as this. Now, what are the great advantages to be gained through such a real obliteration of personality? What is the result for the sake of whose achievements are sacrificed the directness and succinctness such as only the special sense perceptions vouchsafe to physical ideas?

The result is nothing more than the attainment of unity and compactness in our system of theoretical physics, and, in fact, the unity of the system, not only in relation to all of its details, but also in relation to physicists of all places, all times, all peoples, all cultures. Certainly, the system of theoretical physics should be adequate, not only for the inhabitants of this earth, but also for the inhabitants of other heavenly bodies. Whether the inhabitants of liars, in case such actually exist, have eyes and ears like our own, we do not know, ---it is quite improbable; but that they, in so far as they possess the necessary intelligence, recognize the law of gravitation and the principle of energy, most physicists would hold as self evident: and anyone to whom this is not evident had better not appeal to the physicists, for it will always remain for him an unsolvable riddle that the same physics is made in the United States as in Germany.

To sum up, we may say that the characteristic feature of the actual development of the system of theoretical physics is an ever extending emancipation from the anthropomorphic elements, which has for its object the most complete separation possible of the system of physics and the individual personality of the physicist. One may call this the objectiveness of the system of physics. In order to exclude the possibility of any misunderstanding, I wish to emphasize particularly that we have here to do, not with an absolute separation of physics from the physicist---for a physics without the physicist is unthinkable,--- but with the elimination of the individuality of the particular physicist and therefore with the production of a common system of physics for all physicists.

Now, how does this principle agree with the

positivist conceptions mentioned above? Separation of the system of physics from the individual personality of the physicist? Opposed to this principle, in accordance with those conceptions, each particular physicist must have his special system of physics, in case that complete elimination of all metaphysical elements is effected; for physics occupies itself only with the facts discovered through perceptions, and only the individual perceptions are directly involved. That other living beings have sensations is, strictly speaking, but a very probable, though arbitrary, conclusion from analogy. The system of physics is therefore primarily an individual matter and, if two physicists accept the same system, it is a very happy circumstance in connection with their personal relationship, but it is not essentially necessary. One can regard this view-point however he will; in physics it is certainly quite fruitless, and this is all that I care to maintain here. Certainly, I might add, each great physical idea means a further advance toward the emancipation from anthropomorphic ideas. This was true in the passage from the Ptolemaic to the Copernican cosmical system, just as it is true at the present time for the apparently impending passage from the so-called classical mechanics of mass points to the general dynamics originating in the principle of relativity. In accordance with this, man and earth upon which he dwells are removed from the centre of the world. It may be predicted that in this century the idea of time will be divested of the absolute character with which men have been accustomed to endow it (cf. the final lecture). Certainly, the sacrifices demanded by every such revolution in the intuitive point of view are enormous; consequently, the resistance against such a change is very great. But the development of science is not to be permanently halted thereby; on the contrary, its strongest impetus is experienced through precisely those forces which attain success in the struggle against the old points of view, and to this extent such a struggle is constantly necessary and useful.

Now, how far have we advanced today toward the unification of our system of physics? The numerous independent domains of the earlier physics now appear reduced to two; mechanics and electrodynamics, or, as one may. say: the physics of material bodies and the physics of the ether. The former comprehends acoustics, phenomena in material bodies, and chemical phenomena; the latter, magnetism, optics, and radiant heat. But is this division a fundamental one? Will it prove final? This is a question of great consequence for the future development of physics. For myself, I believe it must be answered in the negative, and upon the following grounds: mechanics and electrodynamics cannot be permanently sharply differentiated from each other. Does the process of light emission, for example, belong to mechanics or to electrodynamics? To which domain shall be assigned the laws of motion of electrons? At first glance, one may perhaps say: to electrodynamics, since with the electrons ponderable matter does not play any role. But let one direct his attention to the motion of free electrons in metals. There he will find, in the study of the classical researches of H. A. Lorentz, for example, that the laws obeyed by the electrons belong rather to the kinetic theory of gases than to electrodynamics. In general, it appears to me that the original differences between processes in the ether and processes in material bodies are to be considered as disappearing. Electrodynamics and mechanics are not so remarkably far apart, as is considered to be the case by many people, who already speak of a conflict between the mechanical and the electrodynamic views of the world. Mechanics requires for its foundation essentially nothing more than the ideas of space, of time, and of that which is moving, whether one considers this as a substance or a state. The same ideas are also involved in electrodynamics. A sufficiently generalized conception of mechanics can therefore also well include electrodynamics, and, in fact, there are many indications pointing toward the ultimate amalgamation of these two subjects, the domains of which already overlap in some measure.

If, therefore, the gulf between ether and matter be once bridged, what is the point of view which in the last analysis will best serve in the subdivision of the system of physics? The answer to this question will characterize the whole nature of the further development of our science. It is, therefore, the most important among all those which I propose to treat today. But for the purposes of a closer investigation it is necessary that we go somewhat more deeply into the peculiarities of physical principles.

We shall best begin at that point from which the first step was made toward the actual realization of the unified system of physics previously postulated by the philosophers only; at the principle of conservation of energy. For the idea of energy is the only one besides those of space and time which is common to all the various domains of physics. In accordance with what I have stated above, it will be apparent and quite self evident to you that the principle of energy, before its general formularization by Mayer, Joule, and Helmholz, also bore an anthropomorphic character. The roots of this principle lay already in the recognition of the fact that no one is able to obtain useful work from nothing; and this recognition had originated essentially in the experiences which were gathered in attempts at the solution of a technical problem: the discovery of perpetual motion. To this extent, perpetual motion has come to have for physics a far reaching significance, similar to that of alchemy for the chemist, although it was not the positive, but rather the negative results of these experiments, through which science was advanced. Today we speak of the principle of energy quite without reference to the technical viewpoint or to that of man. We say that the total amount of energy of an isolated system of bodies is a quantity whose amount can be neither increased nor diminished through any kind of process within the system, and we no longer consider the accuracy with which this law holds as dependent upon the refinement of the methods, which we at present possess, of testing experimentally the question of the realization of perpetual motion. In this, strictly speaking, unprovable generalization, impressed upon us with elemental force, lies the emancipation from the anthropomorphic elements mentioned above.

While the principle of energy stands before us as a complete independent structure, freed from and independent of the accidents appertaining to its historical development, this is by no means true in equal measure in the case of that principle which R. Clausius introduced into physics; namely, the second law of thermodynamics. This law plays a very peculiar role in the development of physical science, to the extent that one is not able to assert today- that for it a generally recognized, and there fore objective formularization, has been found. In our present consideration it is therefore a matter of particular interest to examine more closely its significance.

In contrast to the first law of thermodynamics, or the energy principle, the second law maybe characterized as follows. While the first law permits in all processes of nature neither the creation nor destruction of energy, but permits of transformations only, the second law goes still further into the limitation of the possible processes of nature, in that it permits, not all kinds of transformations, but only certain types, subject to certain conditions. The second law occupies itself, therefore, with the question of the kind and, in particular, with the direction of any natural process.

At this point a mistake has frequently been made, which has hindered in a very pronounced manner the advance of science up to the present day. In the endeavor to give to the second law of thermodynamics the most general character possible, it has been proclaimed by followers of W. Ostwald as the second law of energetics, and the attempt made so to formulate it that it shall determine quite generally the direction of every process occurring in nature. Some weeks ago I read in a public academic address of an esteemed colleague the statement that the import of the second law consists in this, that a stone falls downwards, that water flows not up hill, but down, that electricity flows from a higher to a lower potential, and so on. This is a mistake which at present is altogether too prevalent not to warrant mention here.

The truth is, these statements are false. A stone can just as well rise in the air as fall downwards; water can likewise flow upwards, as, for example, in a spring; electricity can flow very well from a lower to a higher potential, as in the case of oscillating discharge of a condenser. The statements are obviously quite correct, if one applies them to a stone originally at rest, to water at rest, to electricity at rest; but then they follow immediately from the energy principle, and one does not need to add a special second law. For, in accordance with the energy principle, the kinetic energy of the stone or of the water can only originate at the cost of gravitational energy, i. e., the center of mass must descend. If, therefore, motion is to take place at all, it is necessary that the gravitational energy shall decrease. That is, the center of mass must descend. In like manner, an electric current between two condenser plates can originate only at the cost of electrical energy already present; the electricity must therefore pass to a lower potential. If, however, motion and current be already present, then one is not able to say, a priori, anything in regard to the direction of the change; it can take place just as well in one direction as the other. Therefore, there is no new insight into nature to be obtained from this point of view.

Upon an equally inadequate basis rests another conception of the second law, which I shall now mention. In considering the circumstance that mechanical work may very easily be transformed into heat, as by friction, while on the other hand heat can only with difficulty be transformed into work, the attempt has been made so to characterize the second law, that in nature the transformation of work into heat can take place completely, while that of heat into work, on the other hand, only incompletely and in such manner that every- time a quantity of heat is transformed into work another corresponding quantity of energy must necessarily undergo at the same time a compensating transformation, as, e. g., the passage of heat from a higher to a lower temperature. This assertion is in certain special cases correct, but does not strike in general at the true import of the matter, as I shall show by a simple example.

One of the most important laws of ther nodynamics is, that the total energy of an ideal gas depends only upon its temperature, and not upon its volume. If an ideal gas be allowed to expand while doing work, and if the cooling of the gas be prevented through the simultaneous addition of heat from a heat reservoir at higher temperature, the gas remains unchanged in temperature and energy content, and one may say that the heat furnished by the heat reservoir is completely transformed into work without exchange of energy. Not the least objection can be urged against this assertion. The law of incomplete transformation of heat into work is retained only through the adoption of a different point of view, but which has nothing to do with the status of the physical facts and only modifies the way of looking at the matter, and therefore can neither be supported nor contradicted through facts; namely, through the introduction ad hoc of new particular kinds of energy, in that one divides the energy of the gas into numerous parts which individually can depend upon the volume. But it is a priori evident that one can never derive from so artificial a definition a new physical law, and it is with such that we have to do when we pass from the first law, the principle of conservation of energy, to the second law.

I desire now to introduce such a new physical law: "It is not possible to construct a periodically functioning motor which in principle does not involve more than the raising of a load and the cooling of a heat reservoir." It is to be understood, that in one cycle of the motor quite arbitrary complicated processes may take place, but that after the completion of one cycle there shall remain no other changes in the surroundings than that the heat reservoir is cooled and that the load is raised a corresponding distance, which may be calculated from the first law. Such a motor could of course be used at the same time as a refrigerating machine also, without any further expenditure of energy and materials. Such a motor would moreover be the most efficient in the world, since it would involve no cost to run it; for the earth, the atmosphere, or the ocean could be utilized as the heat reservoir. We shall call this, in accordance with the proposal of W. Ostwald, perpetual motion of the second kind. Whether in nature such a motion is actually possible cannot be inferred from the energy principle, and may only be determined by special experiments.

Just as the impossibility of perpetual motion of the first kind leads to the principle of the conservation of energy, the quite independent principle of the impossibility of perpetual motion of the second kind leads to the second law of thermodynamics, and, if we assume this impossibility as proven experimentally, the general law follows immediately: *there are processes in nature which in no possible way can be made completely reversible.* For consider, e. g., a frictional process through which mechanical work is transformed into heat with the aid of suitable apparatus, if it were actually possible to make in some way such complicated apparatus completely reversible, so that everywhere in nature exactly the same conditions be reestablished as existed at the beginning of the frictional process, then the apparatus considered would be nothing more than the motor described above, furnishing a perpetual motion of the second kind. This appears evident immediately, if one clearly perceives what the apparatus would accomplish: transformation of heat into work without any further outstanding change.

We call such a process, which in no wise can be made completely reversible, an irreversible process, and all other processes reversible processes; and thus we strike the kernel of the second law of thermodynamics when we say that irreversible processes occur in nature. In accordance with this, the changes in nature have a unidirectional tendency. With each irreversible process the world takes a step forward, the traces of which under no circumstances can be completely obliterated. Besides friction, examples of irreversible processes are: heat conduction, diffusion, conduction of electricity in conductors of finite resistance, emission of light and heat radiation, disintegration of the atom in radioactive substances, and so on. On the other hand, examples of reversible processes are: motion of the planets, free fall in empty space, the undamped motion of a pendulum, the frictionless flow of liquids, the propagation of light and sound waves without absorption and refraction, undamped electrical vibrations, and so on. For all these processes are already periodic or may be made completely reversible through suitable contrivances, so that there remains no outstanding change in nature; for example, the free fall of a body whereby the acquired velocity is utilized to raise the body again to its original height; a light or sound wave which is allowed in a suitable manner to be totally reflected from a perfect mirror.

What now are the general properties and criteria of irreversible processes, and what is the general quantitative measure of irreversibility? This question has been examined and answered in the most widely different ways, and it is evident here again how difficult it is to reach a correct formularization of a problem. Just as originally we came upon the trail of the energy principle through the technical problem of perpetual motion, so again a technical problem, namely, that of the steam engine, led to the differentiation between reversible and irreversible processes. Long ago Sadi Carnot recognized, although he utilized an incorrect conception of the nature of heat, that irreversible processes are less economical than reversible, or that in an irreversible process a certain opportunity to derive mechanical work from heat is lost. What then could have been simpler than the thought of making, quite in general, the measure of the irreversibility of a process the quantity of mechanical work which is unavoidably lost in the process. For a reversible process then, the unavoidably lost work is naturally to be set equal to zero. This view, in accordance with which the import of the second law consists in a dissipation of useful energy, has in fact, in certain special cases, e. g., in isothermal processes, proved itself useful. It has persisted, therefore, in certain of its aspects up to the present day; but for the general case, however, it has shown itself as fruitless and, in fact, misleading. The reason for this lies in the fact that the question concerning the lost work in a given irreversible process is by no means to be answered in a determinate manner, so long as nothing further is specified with regard to the source of energy from which the work considered shall be obtained.

An example will make this clear. Heat conduction is an irreversible process, or as Clausius expresses it: Heat cannot without compensation pass from a colder to a warmer body. What now is the work which in accordance with definition is lost when the quantity of heat Q passes through direct conduction from a warmer body at the temperature T_1 to a colder body at the temperature T_2? In order to answer this question, we make use of the heat transfer involved in carrying out a reversible Carnot cyclical process between the two bodies employed as heat reservoirs. In this process a certain amount of work would be obtained, and it is just the amount sought, since it is that which would be lost in the direct passage by conduction; but this has no definite value so long as we do not know whence the work originates, whether, e. g., in the warmer body or in the colder body, or from somewhere else. Let one reflect that the heat given up by the warmer body in the reversible process is certainly not equal to the heat absorbed by the colder body, because a certain amount of heat is transformed into work, and that we can identify, with exactly the same right, the quantity of heat Q transferred by the direct process of conduction with that which in the cyclical process is given up by the warmer body, or with that absorbed by the colder body. As one does the former or the latter, he accordingly obtains for the quantity of lost work in the process of conduction:

$$Q \cdot \frac{T_1 - T_2}{T_1} \quad \text{or} \quad Q \cdot \frac{T_1 - T_2}{T_2}.$$

We see, therefore, that the proposed method of expressing mathematically the irreversibility of a process does not in general effect its object, and at the same time we recognize the peculiar reason which prevents its doing so. The statement of the question is too anthropomorphic. It is primarily too much concerned with the needs of mankind, in that it refers directly to the acquirement of useful work. If one require from nature a determinate answer, he must take a more general point of view, more disinterested, less economic. We shall now seek to do this.

Let us consider any typical process occurring in nature. This will carry all bodies concerned in it from a determinate initial state, which I designate as state A, into a determinate final state B The process is either reversible or irreversible. A third possibility is excluded. But whether it is reversible or irreversible depends solely upon the nature of the two states A and B, and not at all upon the way in which the process has been carried out; for we are only concerned with the answer to the question as to whether or not, when the state B is once reached, a complete return to A in any conceivable manner may be accomplished. If now, the complete return from B to A is not possible, and the process therefore irreversible, it is obvious that the state B may be distinguished in nature through a certain property from state A. Several years ago I ventured to express this as follows: that nature possesses a greater "preference" for state B than for state A In accordance with this mode of expression, all those processes of nature are impossible for whose final state nature possesses a smaller preference than for the original state. Reversible processes constitute a limiting case; for such, nature possesses an equal preference for the initial and for the final state, and the passage between them takes place as well in one direction as the other.

We have now to seek a physical quantity whose magnitude shall serve as a general measure of the preference of nature for a given state. This quantity must be one which is directly determined by the state of the system considered, without reference to the previous history of the system, as is the case with the energy, with the volume, and with other properties of the system. It should possess the peculiarity of increasing in all irreversible processes and of remaining unchanged in all reversible processes, and the amount of change which it experiences in a process would furnish a general measure for the irreversibility of the process.

R. Clausius actually found this quantity and called it "entropy." Every system of bodies possesses in each of its states a definite entropy, and this entropy expresses the preference of nature for the state in question. It can, in all the processes which take place within the system, only increase and never decrease. If it be desired to consider a process in which external actions upon the system are present, it is necessary to consider those bodies in which these actions originate as constituting part of the system; then the law as stated in the above form is valid. In accordance with it, the entropy of a system of bodies is simply equal to the sum of the entropies of the individual bodies, and the entropy of a single body is, in accordance with Clausius, found by the aid of a certain reversible process. Conduction of heat to a body increases its entropy, and, in fact, by an amount equal to the ratio of the quantity of heat given the body to its temperature. Simple compression, on the other hand, does not change the entropy.

Returning to the example mentioned above, in which the quantity of heat Q is conducted from a warmer body at the temperature T_1 to a colder body at the temperature T_2, in accordance with what precedes, the entropy of the warmer body decreases in this process, while, on the other hand, that of the colder increases, and the sum of both changes, that is, the change of the total entropy of both bodies, is:

$$-\frac{Q}{T_1}+\frac{Q}{T_2}>0.$$

This positive quantity furnishes, in a manner free from all arbitrary assumptions, the measure of the irreversibility of the process of heat conduction. Such examples may be cited indefinitely. Every chemical process furnishes an increase of entropy.

We shall here consider only the most general case treated by Clausius: an arbitrary reversible or irreversible cyclical process, carried out with any physico-chemical arrangement, utilizing an arbitrary number of heat reservoirs. Since the arrangement at the conclusion of the cyclical process is the same as that at the beginning, the final state of the process is to be distinguished from the initial state solely through the different heat content of the heat reservoirs, and in that a certain amount of mechanical work has been furnished or consumed. Let Q be the heat given up in the course of the process by a heat reservoir at the temperature T, and let A be the total work yielded (consisting, e. g., in the raising of weights); then, in accordance with the first law of thermodynamics:

$$\Sigma Q = A.$$

In accordance with the second law, the sum of the changes in entropy of all the heat reservoirs is positive, or zero. It follows, therefore, since the entropy of a reservoir is decreased by the amount Q/ T through the loss of heat Q that:

$$\sum \frac{Q}{T} \leq 0.$$

This is the well-known inequality of Clausius.

In an isothermal cyclical process, T is the same for all reservoirs. Therefore:

$$\sum Q \leq 0, \quad \text{hence:} \quad A \leq 0.$$

That is: in an isothermal cyclical process, heat is produced and work is consumed. In the limiting case, a reversible isothermal cyclical process, the sign of equality holds, and therefore the work consumed is zero, and also the heat produced. This law plays a leading role in the application of thermodynamics to physical chemistry.

The second law of thermodynamics including all of its consequences has thus led to the principle of increase of entropy. You will now readily understand, having regard to the questions mentioned above, why I express it as my opinion that in the theoretical physics of the future the first and most important differentiation of all physical processes will be into reversible and irreversible processes.

In fact, all reversible processes, whether they take place in material bodies, in the ether, or in both together, show a much greater similarity among themselves than to any irreversible process. In the differential equations of reversible processes the time differential enters only as an even power, corresponding to the circumstance that the sign of time can be reversed. This holds equally well for vibrations of the pendulum, electrical vibrations, acoustic and optical waves, and for motions of mass points or of electrons, if we only exclude every kind of damping. But to such processes also belong those infinitely slow processes of thermodynamics which consist of states of equilibrium in which the time in general plays no role, or, as one may also say, occurs with the zero power, which is to be reckoned as an even power. As Helmholtz has pointed out, all these reversible processes have the common property that they may be completely represented by the principle of least action, which gives a definite answer to all questions concerning any such measurable process, and, to this extent, theory of reversible processes may be regarded as completely established. Reversible processes have, however, the disadvantage that singly and collectively they are only ideal: in actual nature there is no such thing as a reversible process. Every natural process involves in greater or less degree friction or conduction of heat. But in the domain of irreversible processes the principle of least action is no longer sufficient; for the principle of increase of entropy brings into the system of physics a wholly new element, foreign to the action principle, and which demands special mathematical treatment. The unidirectional course of a process in the attainment of a fixed final state is related to it.

I hope the foregoing considerations have sufficed to make clear to you that the distinction between reversible and irreversible processes is much broader than that between mechanical and electrical processes and that, therefore, this difference, with better right than any other, may be taken advantage of in classifying all physical processes, and that it may eventually play in the theoretical physics of the future the principal role.

However, the classification mentioned is in need of quite an essential improvement, for it cannot be denied that in the form set forth, the system of physics is still suffering from a strong dose of anthropomorphism. In the definition of irreversibility, as well as in that of entropy, reference is made to the possibility of carrying out in nature certain changes, and this means, fundamentally, nothing more than that the division of physical processes is made dependent upon the manipulative skill of man in the art of experimentation, which certainly does not always remain at a fixed stage, but is continually being more and more perfected. If, therefore, the distinction between reversible and irreversible processes is actually to have a lasting significance for all times, it must be essentially broadened and made independent of any reference to the capacities of mankind. How this may happen, I desire to state one week from tomorrow. The lecture of tomorrow will be devoted to the problem of bringing before you some of the most important of the great number of practical consequences following from the entropy principle.

Planck's Columbia Lectures

Summary of the First Lecture

Planck began his lecture by expounding on the philosophy of science and the roles of mathematics in comparison to the reality of measurements of scientific phenomena. In his words, "All physical ideas depend upon measurements, and mathematics is the chief tool with which the material is worked". He clearly had misgivings about depending too much on theory, going so far as to call it "dangerous". On the other hand, he added that it is not possible to accurately define the physics of heat solely by human sensations. Only through objectiveness it is possible to advance toward the *unification of our system of physics.*

This unification must encompass various phenomena: acoustics, phenomena related to mechanics, chemical phenomena, magnetism, optics and radiant heat. All of these domains can be reduced to just *mechanics* and *electrodynamics* or the *physics of material bodies* and the *physics of the ether* (he further covers these differences in detail in his last lecture). The amalgamation of *ether* and *matter* will bridge this gap, and he pointed out that the extent of these two subjects had already overlapped as far back as the early 18th century. Accomplishing this feat was fundamental to his goal, and he was extremely thorough in his investigation, as will be obvious as he proceeds through subsequent lectures.

He contended that the first step in the unification of physics is a philosophical idea; the *principle of conservation of energy.* Planck's three dimensions of the universe are *space*[1], *time* and *energy*, by which all physical phenomena are defined[2]. The conservation of energy had

1 Note that space also has three physical dimensions.

2 Hermann Ludwig Ferdinand von Hemholtz described his *Law of the Conservation of Force* in a series of lectures at Carlsruhe in the

already been formularized by the time of Planck, leading to the confirmation of the ..negative results of perpetual motion experiments. By the *first law of thermodynamics*, energy can only be transformed, not created or destroyed (the energy principle)[3]. In accordance with this law,

$$\Sigma Q = A ,$$

wherein the sum of the heat losses of a reservoir is equal to the total work yielded.

The second law of thermodynamics (Clausius) imposes further limitations to the first law, allowing only certain types of transformations subject to certain conditions. In accordance with this law,

$$\sum \frac{Q}{T} \leq 0,$$

the total decrease of entropy through the loss of heat is either positive or zero. In an isothermal cyclical process, the temperature T is the same for all bodies, and therefore

$$\sum Q \leq 0, \quad \text{hence: } A \leq 0.$$

Heat is produced and work is consumed, and in the limiting case, the work consumed is zero, the heat produced is zero, and the equality holds.

Planck determined the relationships of the concepts of *reversibility* and *irreversibility* in relationship to the

---

winter of 1862-1863. The conservation of energy and the conservation of force are directly related.

3 Some modern physicists have challenged this law, contending that matter can be created from nothing.

second law. An example of reversibility is the lossless charging or discharge of a capacitor or inductor, while the resulting heat loss by the resistance in the circuit is typical of an irreversible process. Perpetual motion exists only in theory for a lossless system, such as an electrical system that includes a coil or capacitor, since components without resistance cannot be obtained, nor can the heat loss be recaptured.

Planck thus established certain basic principles and characterized them mathematically, firstly by the quantity of heat lost by one body to another as a function of the temperatures of the two bodies. A physical body begins in an initial state and evolves into a final state, and if the reverse process occurs exactly, then the process is reversible; if not, then it is irreversible. A frictional process that transforms mechanical energy into work is an irreversible process, since heat is lost. If this were not the case, then perpetual motion would result. The importance of the principle of reversibility with respect to *molecular states* is described in later chapters.

Clausius was credited with defining the irreversibility of a process by a quantity called *entropy*, which is a function of the quantity of heat exchanged in going from one state to another at a given temperature ($\Delta S = -Q/T$). This statement is in accordance with the first law of thermodynamics, which is the principle of the conservation of energy. Therefore, the first body loses a certain quantity of heat in the process, so the entropy of the first body always exceeds that of the second body in the process for an isothermal system. Consequently, the total entropy of the two bodies is always positive, since the entropy of the first body is always greater than that of the second body that receives the lesser energy. This definition describes an irreversible process wherein the possibility of perpetual motion is refuted by the second law of thermodynamics. Details of Planck's explanation of this subject cover the relationships between the conservation of energy, perpetual motion, heat transfer, irreversibility and why entropy always increases (except for certain specific instances).

For the limiting case, a reversible isothermal cyclical process, no work is consumed and no heat is produced, and therefore there is no change in entropy. Examples of reversible processes are the undamped vibrations of the pendulum, electrical and acoustic vibrations and electrons, in which there is no heat loss. All reversible processes have the common property of being in accordance the Principle of Least Action (see footnote 2 above and Chapter 8).

Planck asserted that, in the future, the most important differentiation of physical processes would be into reversible and irreversible processes. Reversible processes show a much greater similarity among themselves than irreversible processes. The definitions of reversibility and irreversibility go beyond the distinctions between electrical and mechanical processes. They are time dependent, and according to Helmholtz have the common property of being completely represented by the Principle of Least Action. The significance of this principle is described by the theoretical methods presented in the seventh lecture of this series.

Throughout these lectures, Planck comments regarding the presence of reversible processes in nature may be somewhat confusing. In this lecture, he states "...in actual nature there is no such thing as reversible processes". In later lectures he seems to contradict this conclusion. What he is most likely referring to is the impossibility of observing reversible processes, which may be part of the anthropomorphic problem that was discussed earlier. Reversible processes can be demonstrated in various forms, such as for the electronic oscillator. There is evidence of the presence of these cyclical processes in nature which are fundamental to his theory and which he examines in detail in the chapters that follow.

In his extraordinary second lecture, Planck elaborates further on the entropy principle and applies it to the analysis of measurable phenomena. His investigative methods are extensive, thorough and clearly described, resulting in a simplicity and beauty that is remarkable.

Planck's Columbia Lectures

Chapter 2

Introduction to the Second Lecture:

In his first lecture, Planck said that he did not intend to go into details, but only to make the general laws of physics more clear by means of a few chosen examples. However, in this lecture he does go into details, describing the physical relationships of the many of the laws of physics and chemistry to real world measurements. Those who are interested in the realities of science will no doubt be fascinated by the scientific methodology he employs.

Planck begins by extending the definition of entropy to include mathematical differentials and their relationship to the laws of thermodynamics. In the next seven pages, develops seven basic equations that are fundamental to the methods in this analysis. Of these seven equations, only two are utilized in all of the seven applications that are analyzed. The results that he obtains with these few equations are quite amazing. The simplicity, efficiency and beauty by which he relates the various laws of physics to real world experiences is the work of a master scientist.

There is a great deal of useful and interesting information in this chapter. Science teachers (including chemists and engineers) could very well draw from it for a course in thermodynamics. To fully comprehend all of this content could take several readings, especially for young science students. The commentary at the end of this chapter may be easier reading, and the details of the interesting applications have not been abbreviated to any significant degree.

SECOND LECTURE.

THERMODYNAMIC STATES OF EQUILIBRIUM IN DILUTE SOLUTIONS.

In the lecture of yesterday I sought to make clear the fact that the essential, and therefore the final division of all processes occurring in nature, is into reversible and irreversible processes, and the characteristic difference between these two kinds of processes, as I have further separated them, is that in irreversible processes the entropy increases, while in all reversible processes it remains constant. Today I am constrained to speak of some of the consequences of this law which will illustrate its rich fruitfulness. They have to do with the question of the laws of thermodynamic equilibrium. Since in nature the entropy can only increase, it follows that the state of a physical configuration which is completely isolated, and in which the entropy of the system possesses absolute maximum, is necessarily a state of stable equilibrium, since for it no further change is possible. How deeply this law underlies all physical and chemical relations has been shown by no one better and more completely than by John Willard Gibbs, whose name, not only in America, but in the whole world will be counted among those of the most famous theoretical physicists of all times; to whom, to my sorrow, it is no longer possible for me to tender personally my respects. It would be gratuitous for me, here in the land of his activity, to expatiate fully on the progress of his ideas, but you will perhaps permit me to speak in the lecture of today of some of the important applications in which thermodynamic research, based on Gibbs works, can be advanced beyond his results. These applications refer to the theory of dilute solutions, and we shall occupy ourselves today with these, while I show you by a definite example what fruitfulness is inherent in thermodynamic theory. I shall first characterize the problem quite generally. It has to do with the state of equilibrium of a material system of any number of arbitrary constituents in an arbitrary number of phases, at a given temperature T and given pressure p. If the system is completely isolated, and therefore guarded against all external thermal and mechanical actions, then in any ensuing change the entropy of the system will increase:

$$dS > 0.$$

But if, as we assume, the system stands in such relation to its surroundings that in any change which the system undergoes the temperature T and the pressure p are maintained constant, as, for instance, through its introduction into a calorimeter of great heat capacity and through loading with a piston of fixed weight, the inequality would suffer a change thereby. We must then take account of the fact that the surrounding bodies also, e. g., the calorimetric liquid, will be involved in the change. If we denote the entropy of the surrounding bodies by S_0, then the following more general equation holds:

$$dS + dS_0 > 0.$$

In this equation

$$dS_0 = -\frac{Q}{T},$$

if Q denote the heat which is given up in the change by the surroundings to the system. On the other hand, if U denote the energy, V the volume of the system, then, in accordance with the first law of thermodynamics,

$$Q = dU + pdV.$$

Consequently, through substitution:

$$dS - \frac{dU + pdV}{T} > 0$$

or, since p and T are constant:

$$d\left(S - \frac{U + pV}{T}\right) > 0.$$

If, therefore, we put:

$$S - \frac{U + pV}{T} = \Phi, \qquad (1)$$

Then

$$d\Phi > 0,$$

and we have the general law, that in every isothermal-isobaric (T = const., p = const.) change of state of a physical system the quantity Φ increases. The absolutely stable state of equilibrium of the system is therefore characterized through the maximum of Φ:

$$\partial\Phi = 0. \qquad (2)$$

If the system consist of numerous phases, then, because Φ, in accordance with (1), is linear and homogeneous in S, U and V, the quantity Φ referring to the whole system is the sum of the quantities Φ referring to the individual phases. If the expression for Φ is known as a function of the independent variables for each phase of the system, then, from equation (2), all questions concerning the conditions of stable equilibrium may be answered. Now, within limits, this is the case for dilute solutions. By "solution" in thermodynamics is meant each homogeneous phase, in whatever state of aggregation, which is composed of a series of different molecular complexes, each of which is represented by a definite molecular number. If the molecular number of a given complex is great with reference to all the remaining complexes, then the solution is called dilute, and the molecular complex in question is called the solvent; the remaining complexes are called the dissolved substances.

Let us now consider a dilute solution whose state is determined by the pressure p, the temperature T, and the molecular numbers $n_0, n_1, n_2, n_3, \ldots$, wherein the subscript zero refers to the solvent. Then the numbers $n_1, n_2, n_3, \ldots$ are all small with respect to n_0, and on this account the volume V and the energy U are linear functions of the molecular numbers:

$$V = n_0 v_0 + n_1 v_1 + n_2 v_2 + \cdots,$$
$$U = n_0 u_0 + n_1 u_1 + n_2 u_2 + \cdots,$$

wherein the v's and u's depend upon p and T only.

From the general equation of entropy:

$$dS = \frac{dU + pdV}{T},$$

in which the differentials depend only upon changes in p and T, and not in the molecular numbers, there results therefore:

$$dS = n_0 \frac{du_0 + pdv_0}{T} + n_1 \frac{du_1 + pdv_1}{T} + \cdots,$$

and from this it follows that the expressions multiplied by $n_0, n_1 \ldots,$ dependent upon p and T only, are complete differentials. We may therefore write:

$$\frac{du_0 + pdv_0}{T} = ds_0\,, \quad \frac{du_1 + pdv_1}{T} = ds_1\,, \quad \cdots \qquad (3)$$

and by integration obtain:

$$S = n_0 s_0 + n_1 s_1 + n_2 s_2 + \cdots + C\,.$$

The constant C of integration does not depend upon p and T, but may depend upon the molecular numbers n_0, n_1, n_2, $\cdots$. In order to express this dependence generally, it suffices to know it for a special case, for fixed values of p and T. Now every solution passes, through appropriate increase of temperature and decrease of pressure, into the state of a mixture of ideal gases, and for this case the entropy is fully known, the integration constant being, in accordance with Gibbs:

$$C = - R(n_0 \log c_0 + n_1 \log c_1 + \ldots),$$

Wherein R denotes the absolute gas constant and c_0, c_1, c_2, $\cdots$ denote the "molecular concentrations":

$$c_0 = \frac{n_0}{n_0 + n_1 + n_2 + \cdots}, \quad c_1 = \frac{n_1}{n_0 + n_1 + n_2 + \cdots}, \quad \cdots.$$

Consequently, quite in general, the entropy of a dilute solution is:

$$S = n_0(s_0 - R \log c_0) + n_1(s_1 - R \log c_1) + \cdots,$$

and, finally, from this it follows by substitution in equation (1) that:

$$\Phi = n_0(\varphi_0 - R \log c_0) + n_1(\varphi_1 - R \log c_1) + \cdots, \quad (4)$$

if we put for brevity:

$$\varphi_0 = s_0 - \frac{u_0 + pv_0}{T}, \quad \varphi_1 = s_1 - \frac{u_1 + pv_1}{T}, \cdots \quad (5)$$

all of which quantities depend only upon p and T.

With the aid of the expression obtained for Φ we are enabled through equation (2) to answer the question with regard to thermodynamic equilibrium. We shall first find the general law of equilibrium and then apply it to a series of particularly interesting special cases.

Every material system consisting of an arbitrary number of homogeneous phases may be represented symbolically in the following way:

$$n_0 m_0, n_1 m_1, \cdots \mid n_0' m_0', n_1' m_1', \cdots \mid n_0'' m_0'', n_1'' m_1'', \cdots \mid \cdots$$

Here the molecular numbers are denoted by n, the molecular weights by m, and the individual phases are separated from one another by vertical lines. We shall now suppose that each phase represents a dilute solution. This will be the case when each phase contains only a single molecular complex and therefore represents an absolutely pure substance; for then the concentrations of all the dissolved substances will be zero.

If now an isobaric-isothermal change in the system of such kind is possible that the molecular numbers

$$n_0, n_1, n_2, \ldots, \quad n_0', n_1' n_2', \quad , \quad n_0'', n_1'' n_2'', \cdots$$

change simultaneously by the amounts

$$\delta n_0, \delta n_1, \delta n_2, \cdots, \quad \delta n_0', \delta n_1', \delta n_2', \cdots, \quad \delta n_0'', \delta n_1'', \delta n_2'', \cdots$$

then, in accordance with equation (2), equilibrium obtains with respect to the occurrence of this change if, when T and p are held constant, the function

$$\Phi + \Phi' + \Phi'' + \cdots$$

is a maximum, or, in accordance with equation (4):

$$\Sigma(\varphi_0 - R \log c_0)\delta n_0 + (\varphi_1 - R \log c_1)\delta n_1 + \cdots = 0$$

(the summation Σ being extended over all phases of the system). Since we are only concerned in this equation with the ratios of the δn's, we put

$$\delta n_0 : \delta n_1 : \cdots : \delta n_0' : \delta n_1' \cdots : \delta n_0'' : \delta n_1'' : \ldots = \nu_0 : \nu_1 : \cdots : \nu_0' : \nu_1' : \cdots : \nu_0'' : \nu_1'' : \ldots,$$

wherein we are to understand by the simultaneously changing ν's, in the variation considered, simple integer positive or negative numbers, according as the molecular complex under consideration is formed or disappears in the change. Then the condition for equilibrium is:

$$\sum v_0 \log c_1 + v_1 \log c_1 + \cdots = \frac{1}{R} \sum v_0 \varphi_0 + v_1 \varphi_1 + \cdots = \log K. \quad (6)$$

K and the quantities $\varphi_0, \varphi_1, \varphi_2, \cdots$ depend only upon p and T, and this dependence is to be found from the equations:

$$\frac{\partial \log K}{\partial p} = \frac{1}{R} v_0 \frac{\partial \varphi_0}{\partial p} + v_1 \frac{\partial \varphi_1}{\partial p} + \cdots,$$

$$\frac{\partial \log K}{\partial T} = \frac{1}{R} v_0 \frac{\partial \varphi_0}{\partial T} + v_1 \frac{\partial \varphi_1}{\partial T} + \cdots.$$

Now, in accordance with (5), for any infinitely small change of p and T:

$$d\varphi_0 = ds_0 - \frac{du_0 + p dv_0 + v_0 dp}{T} + \frac{u_0 + p v_0}{T^2} \cdot dT \ ,$$

and consequently, from (3):

$$d\varphi_0 = \frac{u_0 + p \upsilon_0}{T^2} dT - \frac{\upsilon_0 dp}{T},$$

and hence:

$$\frac{\partial \varphi_0}{\partial p} = -\frac{\upsilon_0}{T}, \quad \frac{\partial \varphi_0}{\partial T} = \frac{u_0 + p \upsilon_0}{T^2}.$$

Similar equations hold for the other φ 's and therefore we get:

$$\frac{\partial \log K}{\partial p} = -\frac{1}{RT} \cdot \sum v_0 \upsilon_0 n + v_1 \upsilon_1 + \cdots,$$

$$\frac{\partial \log K}{\partial T} = -\frac{1}{RT^2} \cdot \sum v_0 u_0 + v_2 u_2 + \cdots + p\,(v_0 \upsilon_0 + v_1 u_1 + \cdots)$$

or, more briefly:

$$\frac{\partial \log K}{\partial p} = -\frac{1}{RT} \cdot \Delta V\,, \quad \frac{\partial \log K}{\partial T} = \frac{\Delta Q}{RT^2}\,, \quad (7)$$

if ΔV denote the change in the total volume of the system and ΔQ the heat which is communicated to it from outside, during the isobaric isothermal change considered. We shall now investigate the import of these relations in a series of important applications.

I. *Electrolytic Dissociation of Water.*

The system consists of a single phase:

$$n_0 H_2 0,\ n_1 H^+,\ n_2 H^- 0.$$

The transformation under consideration

$$v_0 : v_1 : v_2 = \delta\, n_0 : \delta\, n_1 : \delta n_2$$

consists in the dissociation of a molecule H_2O into a molecule H^+ and a molecule H^-O, therefore:

$v_0 = -1$, $v_1 = 1$, $v_2 = 1$.

Hence, in accordance with (6), for equilibrium:

$$- \log c_0 + \log c_1 + \log c_2 = \log K,$$

or, since $c_1 = c_2$ and $c_0 = 1$, approximately:

$$2 \log c_1 = \log K.$$

The dependence of the concentration c_1 upon the temperature now follows from (7):

$$2 \frac{\partial \log c_1}{\partial T} = \frac{\Delta Q}{RT^2}.$$

ΔQ, the quantity of heat which it is necessary to supply for the dissociation of a molecule of H_2O into the ions H^+ and H^-0, is, in accordance with Arrhenius, equal to the heat of ionization in the neutralization of a strong univalent base and acid in a dilute aqueous solution, and, therefore, in accordance with the recent measurements of Wormann,[1]

$$\Delta Q = 27{,}857 - 48.5T \text{ gr. cal.}$$

Using the number 1.985 for the ratio of the absolute gas constant R to the mechanical equivalent of heat, it follows that:

$$\frac{\partial \log c_1}{\partial T} = \frac{1}{2 \cdot 1.985} \left(\frac{27{,}857}{T^2} - \frac{48.5}{T} \right),$$

and by integration:

$$\log_{10} c_1 = -\frac{3047.3}{T} - 12.125 \log_{10} T + \text{const.}$$

1 Ad Heydwiller, Ann. d. Phys., 28, 506, 1909

This dependence of the degree of dissociation upon the temperature agrees very well with the measurements of the electric conductivity of water at different temperatures by Kohlrausch and Heydweiller, Noyes, and Lundén.

II. *Dissociation of a Dissolved Electrolyte.*

Let the system consists of an aqueous solution of acetic acid:

$$n_0 H_2O, \quad n_1 H_4C_2O_2, \quad n_2 H^+, \quad n_3 H_3C_2^- O_2.$$

The change under consideration consists in the dissociation of a molecule $H_4C_2O_2$ into its two ions, therefore

$$\nu_0 = 0, \ ;\nu_1 = -1, \quad \nu_2 = 1, \quad \nu_3 = 1$$

Hence, for the state of equilibrium, in accordance with (6):

$$-\log c_1 + \log c_2 + \log c_2 = \log K,$$

or, since $c_2 = c_3$:

$$\frac{c_2^2}{c_1} = K.$$

Now the sum $c_1 + c_2 = c$ is to be regarded as known, since the total number of the undissociated and dissociated acid molecules is independent of the degree of dissociation. Therefore c_1 and c_2 may be calculated from K and c. An experimental test of the equation of equilibrium is possible on account of the connection between the degree of dissociation and electrical conductivity of the solution. In accordance with the electrolytic dissociation theory of Arrhenius, the ratio of the molecular conductivity λ of the solution in any dilution to the molecular conductivity λ of the solution in infinite dilution is:

$$\frac{\lambda}{\lambda_\infty} = \frac{c_2}{c_1 + c_2} = \frac{c_2}{c}$$

since electric conduction is accounted for by the dissociated molecules only. It follows then, with the aid of the last equation, that:

$$\frac{\lambda^2 c}{\lambda_\infty - \lambda} = K \cdot \lambda_\infty = \text{const.}$$

With unlimited decreasing c, λ increases to λ_∞. This "law of dilution" for binary electrolytes, first enunciated by Ostwald, has been confirmed in numerous cases by experiment, as in the case of acetic acid.

Also, the dependence of the degree of dissociation upon the temperature is indicated here in quite an analogous manner to that in the example considered above, of the dissociation of water.

III. *Vaporization or Solidification of a Pure Liquid.*

In equilibrium the system consists of two phases, one liquid, and one gaseous or solid:

$$n_0 m_0 \mid n_0' m_0'$$

Each phase contains only a single molecular complex (the solvent), but the molecules in both phases do not need to be the same. Now, if a liquid molecule evaporates or solidifies, then in our notation

$$\nu_0 = -1, \quad \nu_0' = \frac{m_0}{m_0'}, \quad c_0 = 1, \quad c_0' = 1,$$

and consequently the condition for equilibrium, in accordance with (6), is:

$$0 = \log K. \qquad (8)$$

Since K depends only upon p and T, this equation therefore expresses a definite relation between p and T: the law of dependence of the pressure of vaporization (or melting pressure) upon the temperature, or vice versa. The import of this law is obtained through the consideration of the dependence of the quantity K upon p and T. If we form the complete differential of the last equation, there results

$$0 = \frac{\partial \log K}{\partial p} dp + \frac{\partial \log K}{\partial T} dT\ ,$$

or, in accordance with (7):

$$0 = -\frac{\Delta V}{T} dp + \frac{\Delta Q}{T^2} dT.$$

If v_0 and v_0' denote the molecular volumes of the two phases, then

$$\Delta V = \frac{m_0 v_0'}{m_0'} - v_0 ,$$

consequently:

$$\Delta Q = T(\frac{m_0 v_0'}{m_0'} - v_0) \frac{dp}{dT} ,$$

or, referred to unit mass:

$$\frac{\Delta Q}{m_0} = T(\frac{v_0'}{m_0'} - \frac{v_0}{m_0}) \cdot \frac{dp}{dT} ,$$

the well-known formula of Carnot and Clapeyron.

IV. *The Vaporization or Solidification of a Solution of Non-Volatile Substances.*

Most aqueous salt solutions afford examples. The symbol of the system in this case is, since the second phase (gaseous or solid) contains only a single molecular complex:

$$n_0 m_0,\ n_1 m_1,\ n_2 m_2, \cdots \mid n_0' m_0' .$$

The change is represented by:

$$\nu_0 = -1, \quad \nu_1 = 0, \quad \nu_2 = 0, \cdots \nu_0' = \frac{m_0}{m_0'} ,$$

and hence the condition of equilibrium, in accordance with (6), is;

$$-\log c_0 = \log K.$$

or, since to small quantities of higher order:

$$c_0 = \frac{n_0}{n_0 + n_1 + n_2 + \ldots} = 1 - \frac{n_1 + n_2 + \ldots}{n_0},$$

$$\frac{n_1 + n_2 + \cdots}{n_0} = K. \qquad (9)$$

A comparison with formula (8), found in example III, shows that through the solution of a foreign substance there is involved in the total concentration a small proportionate departure from the law of vaporization or solidification which holds for the pure solvent. One can express this, either by saying: at a fixed pressure p, the boiling point or the freezing point T of the solution is different than that (T_0) for the pure solvent, or: at a fixed pressure T the vapor pressure or solidification pressure p of the solution is different from that (p_0) of the pure solvent. Let us calculate the departure in both cases:

1. If T_0 be the boiling (or freezing temperature) of the pure solvent at the pressure p, then, in accordance with (8):

$$(\log K)_{T=T_0} = 0,$$

and by subtraction of (9) there results:

$$\log K - (\log K)_{T=T_0} = \frac{n_1 + n_2 + \cdots}{n_0},$$

Now, since T is little different from T_0, we may write in place of this equation, with the aid of (7)

$$\frac{\partial \log K}{\partial T}(T - T_0) = \frac{\Delta Q}{RT_0^2}(T - T_0) = \frac{n_1 + n_2 + \cdots}{n_0},$$

and from this it follows that:

$$(T - T_0) = \frac{n_1 + n_2 + \cdots}{n_0} \cdot \frac{RT_0^2}{\Delta Q} \qquad (10)$$

This is the law for the raising of the boiling point or for the lowering of the freezing point, first derived by van't Hoff: in the case of freezing ΔQ (the heat taken from the surroundings during the freezing of a liquid molecule) is negative. Since n_0 and ΔQ occur only as a product, it is not possible to infer anything from this formula with regard to the molecular number of the liquid solvent.

2. If p_0 be the vapor pressure of the pure solvent at the temperature T, then, in accordance with (8):

$$(\log K)_{p=p_0} = 0,$$

and by subtraction of (9) there results:

$$\log K - (\log K)_{p=p_0} = \frac{n_1 + n_2 + \cdots}{n_0}.$$

Now, since p and p_0 are nearly equal, with the aid of (7) we may write:

$$\frac{\partial \log K}{\partial p}(p - p_0) = -\frac{\Delta V}{RT}(p - p_0) = \frac{n_1 + n_2 + \cdots}{n_0},$$

and from this it follows, if ΔV be placed equal to the volume of the gaseous molecule produced in the vaporization of a liquid molecule:

$$\Delta V = \frac{m_0}{m_0'}\frac{RT}{p},$$

$$\frac{p_0 - p}{p} = \frac{m_0'}{m_0}\cdot\frac{n_1 + n_2 + \cdots}{n_0}.$$

This is the law of relative depression of the vapor pressure, first derived by van't Hoff. Since n_0 and m_0 occur only as a product, it is not possible to infer from this formula anything with regard to the molecular weight of the liquid solvent. Frequently the factor m_0'/m_0 is left out in this formula; but this is not allowable when m_0 and m_0' are unequal (as, e. g:, in the case of water).

V. *Vaporization of a Solution of Volatile Substances. (E. g., a Sufficiently Dilute Solution of Propyl Alcohol in Water.)*

The system, consisting of two phases, is represented by the following symbol:

$$n_0 m_0,\ n_1 m_1,\ n_2 m_2,\quad |\ n_0' m_0',\ n_1' m_1',\ n_2' m_2',\quad ,$$

wherein, as above, the figure 0 refers to the solvent and the figures 1, 2, 3 . . . refer to the various molecular complexes of the dissolved substances. By the addition of primes in the case of the molecular weights (m_0', m_1', m_2' . . .) the possibility is left open that the various molecular complexes in the vapor may possess a different molecular weight than in the liquid.

Since the system here considered may experience various sorts of changes, there are also various conditions of equilibrium to fulfill, each of which relates to a definite sort of transformation. Let us consider first that change which consists in the vaporization of the solvent. In accordance with our scheme of notation, the following conditions hold:

$$\nu_0 = -1, \nu_1 = 0, \nu_2 = 0, \cdots \nu_0{}' = \frac{m_0}{m_0{}'}, \nu_1{}' = 0, \nu_2{}' = 0, \cdots,$$

and, therefore, the condition of equilibrium (6) becomes:

$$-\log c_0 + \frac{m_0}{m_0{}'} \log c_0{}' = \log K,$$

or, if one substitutes:

$$c_0 = 1 - \frac{n_1 + n_2 + \cdots}{n_0} \quad \text{and} \quad c_0{}' = 1 - \frac{n_1{}' + n_2{}' + \cdots}{n_0{}'},$$

$$\frac{n_1 + n_2 + \cdots}{n_0} - \frac{m_0}{m_0{}'} \cdot \frac{n_1{}' + n_2{}' + \cdots}{n_0{}'} = \log K.$$

If we treat this equation upon equation (9) as a model, there results an equation similar to (10):

$$T - T_0 = \left(\frac{n_1 + n_2 + \cdots}{n_0} - \frac{n_1{}' + n_2{}' + \cdots}{n_0{}'} \right) = \log K.$$

Here ΔQ is the heat effect in the vaporization of one molecule of the solvent and, therefore, $\Delta Q/m_0$ is the heat effect in the vaporization of a unit mass of the solvent.

We remark, once more, that the solvent always occurs in the formula through the mass only, and not through the molecular number or the molecular weight, while, on the other hand, in the case of the dissolved substances, the molecular state is characteristic on account of their influence upon vaporization. Finally, the formula contains a generalization of the law of van't Hoff, stated above, for the raising of the boiling point, in that here in place of the number of dissolved molecules in the liquid, the difference between the number of dissolved molecules in unit mass of the liquid and in unit mass of the vapor appears. According as the unit mass of liquid or the unit mass of vapor contains more dissolved molecules, there results for the solution a raising or lowering of the boiling point; in the limiting case, when both quantities are equal, and the mixture therefore boils without changing, the change in boiling point becomes equal to zero. Of course, there are corresponding laws holding for the change in the vapor pressure.

Let us consider now a change which consists in the vaporization of a dissolved molecule. For this case we have in our notation

$$\nu_0 = 0, \nu_1 = -1, \nu_2 = 0 \cdots, \nu_0' = 0, \nu_1' = \frac{m_1}{m_1'}, \nu_2' = 0, \cdots$$

and, in accordance with (6), for the condition of equilibrium:

$$-\log c_1 + \frac{m_1}{m_1'} \log c_1' = \log K$$

or:

$$\frac{c_1'^{\frac{m_1}{m_1'}}}{c_1} = K.$$

This equation expresses the Nernst law of distribution. If the dissolved substance possesses in both phases the same molecular weight ($m_1 = m_1'$), then, in a state of equilibrium a fixed ratio of the concentrations c_1 and c_1' in the liquid and in the vapor exists, which depends only upon the pressure and temperature. But, if the dissolved substance polymerises somewhat in the liquid, then the relation demanded in the last equation appears in place of the simple ratio.

VI. *The Dissolved Substance only Passes over into the Second Phase.*

This case is in a certain sense a special case of the one preceding. To it belongs that of the solubility of a slightly soluble salt, first investigated by van't Hoff, e. g., succinic acid in water. The symbol of this system is:

$$n_0H_20,\ n_1H_6C_4O_4 \mid n_0'\ H_6C_4O_4,$$

in which we disregard the small dissociation of the acid solution. The concentrations of the individual molecular complexes are:

$$c_0 = \frac{n_0}{n_0 + n_1},\quad c_1 = \frac{n_1}{n_0 + n_1},\quad c_0' = \frac{n_0'}{n_0'} = 1$$

For the precipitation of solid succinic acid we have:

$$\nu_0 = 0,\quad \nu_1 = -1,\quad \nu_0' = 1\ ,$$

and, therefore, from the condition of equilibrium (6):

$$-\log c_1 = \log K,$$

hence, from (7)

$$\Delta Q = -R T^2 \frac{\partial \log c_1}{\partial T}.$$

By means of this equation van't Hoff calculated the heat of solution ΔQ from the solubility of succinic acid at 0° and at 8:5 °C. The corresponding numbers were 2.88 and 4.22 in an arbitrary unit. Approximately, then:

$$\frac{\partial \log c_1}{\partial T} = \frac{\log_e 4.22 - \log_e 2.88}{8.5} = 0.04494,$$

from which for $T = 273$:

$$\Delta Q = -1.98 \cdot 273^2 . 0.04494 = -6{,}600 \text{ cal.},$$

that is, in the precipitation of a molecule of succinic acid, 6,600 cal. are given out to the surroundings. Berthelot found, however, through direct measurement, 6,700 calories for the heat of solution.

The absorption of a gas also comes under this head, e. g. carbonic acid, in a liquid of relatively unnoticeable smaller vapor pressure, e. g., water at not too high a temperature. The symbol of the system is then

$$n_0 H_2 0,\ n_1 CO_2 \mid n_0' CO_2.$$

The vaporization of a molecule CO_2 corresponds to the

values

$$\nu_0 = 0, \quad \nu_1 = -1, \quad \nu_0' = 1.$$

The condition of equilibrium is therefore again:

$$-\log c_1 = \log K,$$

i. e., at a fixed temperature and a fixed pressure the concentration c_1 of the gas in the solution is constant. The change of the concentration with p and T is obtained through substitution in equation (7). It follows from this that:

$$\frac{\partial \log c_1}{\partial p} = \frac{\Delta V}{RT}, \quad \frac{\partial \log c_1}{\partial T} = -\frac{\Delta Q}{RT^2}.$$

ΔV is the change in volume of the system which occurs in the isobaric-isothermal vaporization of a molecule of CO_2, ΔQ the quantity of heat absorbed in the process from outside. Now, since ΔV represents approximately the volume of a molecule of gaseous carbonic acid, we may put approximately:

$$\Delta V = \frac{RT}{p},$$

and the equation gives:

$$\frac{\partial \log c_1}{\partial p} = \frac{1}{p},$$

which integrated, gives:

$$\log c_1 = \log p + \text{const.}, \qquad c_1 = C \cdot p,$$

i.e., the concentration of the dissolved gas is proportional to the pressure of the free gas above the solution (law of Henry and Bunsen). The factor of proportionality C, which furnishes a measure of the solubility of the gas, depends upon the heat effect in quite the same manner as in the example previously considered.

A number of no less important relations are easily derived as by-products of those found above, e. g., the Nernst laws concerning the influence of solubility, the Arrhenius theory of isohydric solutions, etc. All such may be obtained through the application of the general condition of equilibrium (6). In conclusion, there is one other case that I desire to treat here. In the historical development of the theory this has played a particularly important role

VII. *Osmotic Pressure.*

We consider now a dilute solution separated by a membrane (permeable with regard to the solvent but impermeable as regards the dissolved substance) from the pure solvent (in the same state of aggregation), and inquire as to the condition of equilibrium. The symbol of the system considered we may again take as

$$n_0 m_0, n_1 m_1, n_2 m_2, \cdots \mid n_0' m_0 .$$

The condition of equilibrium is also here again expressed by equation (6), valid for a change of state in which the temperature and the pressure in each phase is maintained constant. The only difference with respect to the cases treated earlier is this, that here, in the presence of a separating membrane between two phases, the pressure p in the first phase may be different from the pressure p' in the second phase, whereby by "pressure," as always, is to be understood the ordinary hydrostatic or manometric pressure.

The proof of the applicability of equation (6) is found in the same way as this equation was derived above, proceeding from the principle of increase of entropy. One has but to remember that, in the somewhat more general case here considered, the external work in a given change is represented by the sum $pdV + p'dV'$, where V and V' denote the volumes of the two individual phases, while before V denoted the total volume of all phases. Accordingly, we use, instead of (7), to express the dependence of the constant K in (6) upon the pressure:

$$\frac{\partial \log K}{\partial p} = -\frac{\Delta V}{RT}, \quad \frac{\partial \log K}{\partial p'} = -\frac{\Delta V'}{RT}. \qquad (11)$$

We have here to do with the following change:

$$\nu_0 = -1, \quad \nu_1 = 0, \quad \nu_2 = 0, \quad \cdots, \quad \nu_0' = 1,$$

whereby is expressed, that a molecule of the solvent passes out of the solution through the membrane into the pure solvent. Hence, in accordance with (6)

$$-\log c_0 = \log K,$$

or, since

$$c_0 = 1 - \frac{n_1 + n_2 + \cdots}{n_0}, \quad \frac{n_1 + n_2 + \cdots}{n_0} = \log K .$$

Here K depends only upon T, p and p'. If a pure solvent were present upon both sides of the membrane, we should have $c_0 = 1$, and $p = p'$; consequently:

$$(\log K)_{p=p'} = 0,$$

and by subtraction of the last two equations:

$$\frac{n_1 + n_2 + \cdots}{n_0} = \log K - (\log K)_{p=p'} = \frac{\partial \log K}{\partial p}(p - p')$$

and in accordance with (11):

$$\frac{n_1 + n_2 + \cdots}{n_0} = -(p - p') \cdot \frac{\Delta V}{RT} .$$

Here ΔV denotes the change in volume of the solution due to the loss of a molecule of the solvent ($\nu_0 = -1$). Approximately then:

$$-\Delta V \cdot n_0 = V ,$$

the volume of the whole solution, and

$$\frac{n_1 + n_2 + \cdots}{n_0} = (p - p') \cdot \frac{\Delta V}{RT} .$$

If we call the difference $p - p'$, the osmotic pressure of the solution, this equation contains the well known law of osmotic pressure, due to van't Hoff.

The equations here derived, which easily permit of multiplication and generalization, have, of course, for the most part not been derived in the ways described above, but have been derived, either directly from experiment, or theoretically from the consideration of special reversible isothermal cycles to which the thermodynamic law was applied, that in such a cyclic process not only the algebraic sum of the work produced and the heat produced, but that also each of these two quantities separately, is equal to zero (first lecture, p. 19). The employment of a cyclic process has the advantage over the procedure here proposed, that in it the connection between the directly measurable quantities and the requirements of the laws of thermodynamics succinctly appears in each case; but for each individual case a satisfactory cyclic process must be imagined, and one has not always the certain assurance that the thermodynamic realization of the cyclic process also actually supplies all the conditions of equilibrium. Furthermore, in the process of calculation certain terms of considerable weight frequently appear as empty ballast, since they disappear at the end in the summation over the individual phases of the process.

On the other hand, the significance of the process here employed consists therein, that the necessary and sufficient conditions of equilibrium for each individually considered case appear collectively in the single equation (6), and that they are derived collectively from it in a direct manner through an unambiguous procedure. The more complicated the systems considered are, the more apparent becomes the advantage of this method, and there is no doubt in my mind that in chemical circles it will be more and more employed, especially, since in general it is now the custom to deal directly with the energies, and not with cyclic processes, in the calculation of heat effects in chemical changes.

Planck's Columbia Lectures

Summary of the Second Lecture

There was a great deal of information presented in this lecture. Planck continued building the foundation of his theory as based on established laws and physical measurements. In each example, he accounted for every molecule of every weight, number and type, all of the pressures and temperatures to which they are subjected, their energy, mass and volume, their gas constants, the energy states that they attain, and their aggregations.

For these applications, Planck chose to utilize certain chemical reactions upon which extensive measurements have been made. Characterizations are made possible by a few restrictions, the use of dilute solutions in these applications being most useful for this purpose. The relationship of the gas constants of liquids and solids and gases to the states of equilibrium is interesting. These enumerated constants are the "parameters" of mathematical equations, which are often generally considered the least interesting terms in a purely mathematical analysis. Clearly, the parameters of equations that represent physical systems can be quite meaningful.

The derivation of the most important seven of the eleven numbered equations that he used in his analysis will be emphasized in this summary. Not all of the mathematical formulas are necessary for comprehension of the methodology. In fact, just two primary equations were utilized in all seven applications.

The state of equilibrium is fundamental to the definition of the molecular states, as are their entropies. A fundamental law of thermodynamics is that an isolated system, whose entropy S reaches a maximum value, is in a state of equilibrium, and he gives credit for this law to the physicist John Willard Gibbs. When a system at a given pressure p and temperature T is completely isolated from external excitation, then the entropy can only increase,

$$dS > 0,$$

which continues until a condition of equilibrium is reached. This relationship applies to a system of any number of constituents in an arbitrary number of phases, which is a very general characterization.

Applying the first law of thermodynamics to an isothermal-isobaric system (T = const., p = const.),

$$dS + dS_0 > 0,$$

in which the entropy of the bodies surrounding the system is denoted by S_0. The change in entropy of the surroundings decreases in the process and is a function of the heat loss Q,

$$dS_0 = -\frac{Q}{T}.$$

Therefore, the change in entropy is proportional to the heat loss and inversely proportional to the absolute temperature.

In accordance with the first law of thermodynamics, the heat loss is function of the energy U, pressure and volume of the system,

$$Q = dU + pdV.$$

In this equation, the amount of heat loss is equal to the change in energy plus the pressure times the change in volume of the system. This differential equation is substituted into the entropy equation,

$$dS - \frac{dU + pdV}{T} > 0,$$

and since p and T are constant,

$$d\left(S - \frac{U + pV}{T}\right) > 0.$$

For every change of state of a physical system (under the conditions stated above), the entropy increases, and therefore the value of the bracketed term in the above equation increases. The energy state Φ of an isothermal-isobaric system is defined as

$$\Phi \triangleq S - \frac{U + pV}{T}, \quad (1)$$

which increases for every change in state,

$$d\Phi > 0.$$

Therefore the system undergoes a series of state changes, and Φ increases until an absolutely stable state of equilibrium is reached and no further change is possible,

$$\delta\,\Phi = 0. \quad (2)$$

Equation (2) represents the differential of Φ alone, regardless of how the other variables change, and the system reaches a stable state when Φ reaches a maximum.

The expression (1) for Φ is a function of the

independent variable for each phase of the system, and equation (2) defines the conditions for stability and equilibrium. Equation (1) is linear and homogeneous in S, U and V, and Φ is the sum of the quantities that represent the phases of the system. If the expression for Φ is known, then equation (2) can be utilized to determine the conditions for stable equilibrium. This is the case for dilute solutions, which is the reason for using this choice in the applications that are analyzed.

In thermodynamics, a "solution" is composed of a series of different molecular complexes, each of which is represented by a molecular number, and one particular molecular number is great with respect to the remaining complexes. The state is determined by the pressure p and the temperature T, as described by the above equation, and the volume V and the energy U are linear functions of the molecular numbers of the constituents:

$$V = n_0 v_0 + n_1 v_1 + n_2 v_2 \cdots = n_0 v_0 + \sum_{k=1}^{N} n_k v_k ,$$

$$U = n_0 u_0 + n_1 u_1 + n_2 u_2 \cdots = n_0 u_0 + \sum_{k=1}^{N} n_k u_k .$$

In these equations, the number of the molecules n_0 in the predominate complex is much greater than the numbers of the molecules in the other complexes, which is the case for a dilute solution. The expression to the far right of the equations is a compact mathematical expression for a series of terms. These expressions encompass a very broad range of possibilities for various dilute solutions.

The next step is to apply the entropy equation. At equilibrium, $\delta\Phi = 0$, and from the general equation of entropy (1)

$$S = \frac{U + pV}{T}.$$

The state is determined by the changes in pressure p and temperature T, and the differential entropy is

$$dS = \frac{dU + pdV}{T},$$

and upon substitution of the above series equations,

$$dS = n_0 \frac{du_0 + pdv_0}{T} T + \sum_{k=1}^{N} n_k \frac{du_k + pdv_k}{T}.$$

Each term in the above equation depends only upon p and T and are complete differentials,

$$\frac{d\,u_0 + p\,dv_0}{T} = ds_0, \quad \frac{d\,u_k + p\,dv_k}{T} = ds_k. \quad (3)$$

Integrating the above series equation term-by-term,

$$S = \sum_{k=0}^{N} n_k s_k + C,$$

in which the individual entropies s_k depend upon p and T, and the constant C may depend only upon the molecular numbers n_k.

The entropy is a definite value when a solution reaches the state where it consists of a mixture of ideal gases after undergoing an appropriate increase of temperature pressure and decrease in pressure. Every solution has this property, and the integration constant is

$$C = -R\sum_{k=0}^{N} n_k \log c_k$$

in accordance with the work of Gibbs. In this equation, R is the absolute gas constant, and c_k denotes the molecular concentrations of the gases,

$$c_k = \frac{n_k}{\sum_k n_k}. .$$

Consequently, the general entropy of a dilute solution is

$$S = \sum_{k=0}^{N} n_k \left(s_k - R\log c_k\right).$$

Substituting the result into equation (1), results in the overall phase equation (4)

$$\Phi = \sum_{k=0}^{N} n_k (\varphi_k - R\log c_k). \quad (4)$$

The gas constant, R, in this equation arises from the fact that every solution passes into a state of a *mixture of ideal gases* as the temperature is increased and the pressure decreased appropriately (per Gibbs).

Equation (5) applies to each of the k individual phases of each molecular complex

$$\varphi_k = s_k - \frac{u_k + p v_k}{T}. \quad (5)$$

in which all quantities depend only upon p and T.

Thus equation (1), which represents the phase state Φ and equation (2), which is the condition for its maximum, were used to determine the expressions for a system in thermodynamic equilibrium. These equations were characterized using the laws for dilute solutions. The next step is to find the general law of equilibrium and apply to a series of special applications.

Planck defined the complex of homogeneous phases of a material system in terms of the individual phases for all of the molecules of each molecular complex,

$$n_0 m_0, n_1 m_1, \cdots, \quad n_0{}' m_0{}', n_1{}' m_1{}', \cdots, n_0{}'' m_0{}'', n_1{}'' m_1{}'', \cdots, \quad \cdots.$$

The n_k are the molecular numbers, and the m_k are the molecular weights. The terms that are primed represent the individual phases. When each phase contains only a single molecular complex, it represents an absolutely pure substance in which case each phase represents a dilute solution, and the concentrations of all of the dissolved substances is zero.

If an isobaric-isothermal change produces a simultaneous change in the molecular numbers, then, in accordance with equation (2) for equilibrium under constant temperature and pressure, the individual phases of the molecular complexes,

is a maximum. Then, in accordance with the above terminology, after first substituting equation (5) into equation (4) results in the change in phase state

$$\sum_{k=0}^{N} (\varphi_k - R \log c_k) \delta n_k = 0,$$

which zero at equilibrium and extends over all phases of the system. The ratios of the δn's in the above equation are proportional to the ratios of the v's,

$$\begin{aligned} &\delta n_0 : \delta n_1, \cdots, \quad \delta n_0' : \delta n_1', \cdots, \quad \delta n_0'' : \delta n_1'', \cdots = \\ &v_0 : v_1, \cdots, \quad v_0' : v_1', \cdots, \quad v_0'' : v_1'', \cdots, \quad . \end{aligned}$$

These ratios are simple integer positive numbers that vary with the molecular complex under consideration, and the condition for equilibrium is

$$\sum v_n \log c_n = \frac{1}{R} \sum v_n \varphi_n = \log K, \quad (6)$$

where c_n are the molecular concentrations, and the individual phases φ_n are functions of pressure p and temperature T.

The final step in this development was to obtain an expression for the variations in K produced by heat exchange and volumetric changes. Differentiating with respect to pressure and then temperature,

$$\frac{\partial \log K}{\partial p} = \frac{1}{R} \sum_{k=0}^{N} v_k \frac{\partial \varphi_k}{\partial p},$$

$$\frac{\partial \log K}{\partial T} = \frac{1}{R} \sum_{k=0}^{N} v_k \frac{\partial \varphi_k}{\partial T}.$$

Then forming the differentials of equation (5) for changes

in p and T and substituting the differentials of equation (3),

$$\frac{du_k + pdv_k}{T} = ds_k,$$

and hence

$$\frac{\partial \varphi_k}{\partial p} = -\frac{v_k}{T}, \quad \frac{\partial \varphi_k}{\partial T} = -\frac{u_k + pv_k}{T^2}.$$

Substituting this result into the above series differential equations for the variations in K with p and T,

$$\frac{\partial \log K}{\partial p} = -\frac{1}{RT} \cdot \Delta V, \quad \frac{\partial \log K}{\partial T} = -\frac{\Delta Q}{RT^2}, \quad (7)$$

where ΔV is the total change in the volume of the system and ΔQ is the heat communicated to it from the outside by the isobaric-isothermal exchange in going from one condition of equilibrium to another.

Planck thus derived a set of equations for changes in the state of a system that is based on the principle of entropy and the laws of thermodynamics. These equations were applied to seven practical applications. Of the above seven equations, only two of them [equations (6) and (7)] are used for most of the analysis in the following seven applications.

I. *Electrolytic Dissociation of Water*

In this application, water molecules are split into two ions in a single-phase transformation

$$n_0 H_2 O \rightarrow n_1 H^+ + n_2 H^- O.$$

Therefore, the molecular ratios in equation (6) are simple unit integers, since a water molecule ν_0, is lost by dissociation into two ions (H^+ and H^+O),

$\nu_0 = -1, \qquad \nu_1 = +1 \qquad \nu_2 = +1.$

The equation on the left represents the original water molecule, which is lost by dissociation, into the two ions that are represented by the terms on the right. The resulting approximate molecular concentrations are

$c_0 = 1, \qquad c_1 = c_2.$

Substituting into equation (6),

$$-\log c_0 + \log c_1 + \log c_2 = \log K,$$

and therefore,

$$2 \log c_1 = \log K.$$

The ion concentration c_1 is a function of temperature, as is seen by applying equation (7),

$$2\frac{\partial \log c_1}{\partial T} = \frac{\Delta Q}{RT^2}.$$

The heat of dissociation of a molecule of H_2O into its two ions H^+ and H^-O at a given temperature was determined by measurement (Wormann)

$$\Delta Q = 27{,}857 - 48.5T \text{ gm-cal.}$$

and is in accordance with the heat of ionizaion in the neutralization of a strong univalent base and acid in a dilute aqueous solution (Arrhenius).

Substituting the value of the gas constant into equation (7),

$$\frac{\partial \log c_1}{\partial T} = \frac{1}{2 \cdot 1.985}\left(\frac{27.857}{T^2} - \frac{48.5}{T}\right).$$

Integrating with respect to temperature,

$$\log_{10} c_1 = -\frac{3047.3}{T} - 12.125 \log_{10} T + \text{const.},$$

which agrees with agrees with the measurements of the electric conductivity of water at different temperatures (Kohlrausch and Heydweiller, Noyes and Lunden).

II. Dissociation of a Dissolved Electrolyte.

This application is a bit more complex, involving an aqueous solution of acetic acid that also forms ions, similar to the previous example:

$$n_0 H_2O, \quad n_1 H_4C_2O_2, \quad n_2 H^+, \quad n_3 H_3C_2^- O_2.$$

A single molecule of acetic acid $H_4C_2O_2$ separates into two ions, and the ratios are again unit numbers.

$$\nu_0 = 0, \quad \nu_1 = -1, \quad \nu_2 = 1, \quad \nu_3 = 1$$

(note that the water molecule did not dissociate). These simple numbers are substituted into equation (6), the antilog is taken, and a simple ratio is obtained. At equilibrium, equation (6) becomes:

$$-\log c_1 + \log c_2 + \log c_2 = \log K,$$

and since $c_2 = c_3$,

$$\frac{c_2^{\ 2}}{c_1} = K.$$

The total number of the undissociated and dissociated acid molecules is independent of the degree of dissociation, and so the sum $(c_1 + c_2) = c$ is known . Therefore c_1 and c_2 may be calculated from K and c. Because of the connection between the degree of dissociation and electrical conductivity of the solution, it is possible to test the equation of equilibrium. Applying the electrolytic dissociation theory of Arrhenius, the ratio of the molecular conductivity λ of the solution in any dilution to the molecular conductivity λ of the solution in infinite dilution is:

$$\frac{\lambda}{\lambda_\infty} = \frac{c_2}{c_1 + c_2} = \frac{c_2}{c}$$

(electric conduction is accounted for by the dissociated molecules only). In accordance with the above two equations,

$$\frac{\lambda^{\ 2} c}{\lambda_\infty - \lambda} = K \cdot \lambda_\infty = \text{const.}$$

As c decreases without limit, λ increases to λ_∞. This is Ostwald's "law of dilution" for binary electrolytes, which has been confirmed in numerous cases by experiment, as in the case of acetic acid. The dependence of the degree of dissociation upon the temperature is similar to the example above for the dissociation of water.

III. Vaporization or Solidification of a Pure Liquid.

In this application, a pure liquid either vaporizes or solidifies, resulting in two phases in which the number of molecules in each phase can differ. In equilibrium the system consists of two phases, one liquid, and one gaseous or solid:

$$n_0 m_0 \mid n_0' m_0'.$$

Each phase contains only a single molecular complex.

If a liquid molecule evaporates or solidifies, then

$$v_0 = -1, \quad v_0' = \frac{m_0}{m_0'}, \quad c_0 = 1, \quad c_0' = 1 .$$

The condition for equilibrium, in accordance with (6), is:

$$0 = \log K. \qquad (8)$$

This equation therefore expresses the law of dependence of the pressure of vaporization (or melting pressure) upon the temperature, or vice versa. The quantity K depends only upon p and T, and differentiating with respect to these variables,

$$0 = \frac{\partial \log K}{\partial p} dp + \frac{\partial \log K}{\partial T} dT ,$$

and applying equation (7):

$$0 = -\frac{\Delta V}{T} dp + \frac{\Delta Q}{T^2} dT.$$

For the molecular volumes of the two phases, v_0 and v_0', the change in volume is

$$\Delta V = \frac{m_0 v_0'}{m_0'} - v_0 ,$$

and similarly, the change in heat is:

$$\Delta Q = T(\frac{m_0 v_0'}{m_0'} - v_0) \frac{dp}{dT} ,$$

or, referred to unit mass:

$$\frac{\Delta Q}{m_0} = T(\frac{v_0'}{m_0'} - \frac{v_0}{m_0}) \cdot \frac{dp}{dT} ,$$

which is the well-known formula of Carnot and Clapeyron.

In summary, he used both terms in equation (7) to yield an expression in terms of volume, temperature, pressure and heat and then finds the molecular volumes of the two phases. Upon re-arranging this equation he obtained the "…well-known formula of Carnot and Clapeyron" for the amount of heat change required to separate the molecules into the two phases as a function of temperature and pressure.

IV. The Vaporization or Solidification of a Solution of Non-Volatile Substances.

This application is the vaporization or solidification of a solution of non-volatile substances, such as aqueous salt solutions. The second phase is either gaseous or solid, containing only a single molecular complex. The system parameters are :

$$n_0 m_0,\ n_1 m_1,\ n_2 m_2, \cdots \mid n_0'\ m_0' .$$

Only one of the molecular complexes vaporizes or solidifies, so the ratio numbers of the other complexes are zero, and the change is represented by:

$$\nu_0 = -1, \quad \nu_1 = 0, \quad \nu_2 = 0, \cdots \nu_0' = \frac{m_0}{m_0'},$$

and applying the condition of equilibrium of equation (6),

$$-\log c_0 = \log K .$$

For small quantities of higher order, the molecular concentration is:

$$c_0 = \frac{n_0}{n_0 + n_1 + n_2 + \ldots} = 1 - \frac{n_1 + n_2 + \ldots}{n_0},$$

and the result is simply:

$$\frac{n_1 + n_2 + \cdots}{n_0} = K. \qquad (9)$$

Comparing this result with equation (8) in example III, shows that the solution of a foreign substance results in a small proportionate departure from the law of vaporization or solidification of the pure solvent. At a fixed pressure p, the boiling point or the freezing point T of the solution is different than that (T_0) for the pure solvent, or: at a fixed pressure T the vapor pressure or solidification pressure p of the solution is different from that (p_0) of the pure solvent. The departure in both cases is:

1. If T_0 is the boiling (or freezing temperature) of the pure solvent at the pressure p, then, in accordance with (8):

$$(\log K)_{T=T_0} = 0,$$

and by subtraction of (9):

$$\log K - (\log K)_{T=T_0} = \frac{n_1 + n_2 + \cdots}{n_0},$$

Since T is little different from T_0, with the application of equation (7),

$$\frac{\partial \log K}{\partial T}(T - T_0) = \frac{\Delta Q}{R T_0^2}(T - T_0) = \frac{n_1 + n_2 + \cdots}{n_0},$$

and therefore:

$$(T - T_0) = \frac{n_1 + n_2 + \cdots}{n_0} \cdot \frac{RT_0^2}{\Delta Q} \qquad (10)$$

This is the law for the raising of the boiling point or for the lowering of the freezing point, first derived by van't Hoff.

2. If p_0 is the vapor pressure of the pure solvent at the temperature T, then, in accordance with (8):

$$(\log K)_{p=p_0} = 0,$$

and by subtraction of (9):

$$\log K - (\log K)_{p=p_0} = \frac{n_1 + n_2 + \cdots}{n_0}.$$

Since p and p_0 are nearly equal, applying equation (7),

$$\frac{\partial \log K}{\partial p}(p - p_0) = -\frac{\Delta V}{RT}(p - p_0) = \frac{n_1 + n_2 + \cdots}{n_0}.$$

Therefore, if ΔV is equal to the volume of the gaseous molecule produced in the vaporization of a liquid molecule, then

$$\Delta V = \frac{m_0}{m_0'} \frac{RT}{p},$$

and

$$\frac{p_0 - p}{p} = \frac{m_0'}{m_0} \cdot \frac{n_1 + n_2 + \cdots}{n_0}.$$

This is the law of relative depression of the vapor pressure, first derived by van't Hoff. The factor m_0'/m_0 is frequently left out in this formula, but this is not allowable when m_0 and m_0' are unequal (as, e. g:, in the case of water).

V. Vaporization of a Solution of Volatile Substances.

(E. g., a Sufficiently Dilute Solution of Propyl Alcohol in Water.)

This application is similar to the previous application, except that he now considers the vaporization of a solution of volatile substances, which in this case consists of a dilute solution of propyl alcohol in water. Planck uses the method that he used in application IV in deriving equation (10) from equation (6). He allows for a more complex system, consisting of two phases, which is represented by:

$$n_0 m_0,\ n_1 m_1,\ n_2 m_2, \quad |\ n_0' m_0',\ n_1' m_1',\ n_2' m_2', \quad ,$$

for which the subscript "$_0$" refers to the solvent and the other subscripts $_{1,\,2,\,3}$ . . . refer to the various molecular complexes of the dissolved substances. The addition of primes in the case of the molecular weights (m_0', m_1', m_2' . . .) allows that the various molecular complexes in the vapor may possess different molecular weights.

For a system that may experience various sorts o3f changes, there are also various conditions of equilibrium, each of which relates to a definite type of transformation. The first that change to be considered consists in the vaporization of the solvent. In accordance with our3333 scheme of notation:

$$\nu_0 = -1, \nu_1 = 0, \nu_2 = 0, \cdots \nu_0' = \frac{m_0}{m_0'}, \nu_1' = 0, \nu_2' = 0, \cdots,$$

and, therefore, the condition of equilibrium (6) is:

$$-\log c_0 + \frac{m_0}{m_0'} \log c_0' = \log K,$$

or, by substituting

$$c_0 = 1 - \frac{n_1 + n_2 + \cdots}{n_0} \quad \text{and} \quad c_0' = 1 - \frac{n_1' + n_2' + \cdots}{n_0'}, 33$$

then

$$\frac{n_1 + n_2 + \cdots}{n_0} - \frac{m_0}{m_0'} \cdot \frac{n_1' + n_2' + \cdots}{n_0'} = \log K.$$

Then applying these results to equation (9), results in an equation similar to (10):

$$T - T_0 = \left(\frac{n_1 + n_2 + \cdots}{n_0} - \frac{n_1' + n_2' + \cdots}{n_0'} \right) = \log K.$$

Here ΔQ is the heat effect in the vaporization of one molecule of the solvent and, therefore, $\Delta Q/m_0$ is the heat effect in the vaporization of a unit mass of the solvent.

Here again, the solvent always occurs in the formula through the mass only, and not through the molecular number or the molecular weight. However, in the case of the dissolved substances, the molecular state is characterized bu their influence upon vaporization.

Finally, the formula contains a generalization of

the law of van't Hoff, stated above for the raising of the boiling point. Here the number of dissolved molecules in the liquid is replaced by the difference between the number of dissolved molecules in unit mass of the liquid and in unit mass of the vapor. According as the unit mass of liquid or the unit mass of vapor contains more dissolved molecules, the boiling point of the solution is raised or lowered. In the limiting case, when both quantities are equal, and the mixture therefore boils without changing, the change in boiling point becomes equal to zero. There are corresponding laws holding for the change in the vapor pressure.

In the case of the vaporization of a dissolved molecule,

$$\nu_0 = 0, \nu_1 = -1, \nu_2 = 0 \cdots, \nu_0' = 0, \nu_1' = \frac{m_1}{m_1'}, \nu_2' = 0, \cdots$$

and, in accordance with (6), for the condition of equilibrium:

$$-\log c_1 + \frac{m_1}{m_1'} \log c_1' = \log K,$$

or:

$$\frac{c_1'^{\frac{m_1}{m_1'}}}{c_1} = K.$$

This equation expresses the "Nernst law of distribution". If the dissolved substance possesses the same molecular weight ($m_1 = m_1'$) in both phases, then, in a state of equilibrium a fixed ratio of the concentrations c_1 and c_1' in the liquid and in the vapor exists, which depends only upon the pressure and temperature. However, if the dissolved substance polymerises, to some degree, in the liquid, then the relation demanded in the last equation appears in place of the simple ratio.

Through the use of equations (6) and (7), he was able to determine the heat lost in the precipitation of the succinic acid at room temperature from the heat required at two other temperatures. The results correlate with the measurements of Berthelot. This method is also applicable to the absorption of a gas, and he used CO_2 as and example to derive the concentration of the dissolved gas as a function of the pressure of the free gas above the solution which is in accordance with the law of Henry and Bunsen.

VI. The Dissolved Substance only Passes over into the Second Phase.

This case is in a certain sense a special case of the one preceding. It applies to the solubility of a slightly soluble salt in water, first investigated by van't Hoff, e. g., succinic acid. For this system,

$$n_0H_20,\ n_1H_6C_4O_4 \mid n_0'\ H_6C_4O_4,$$

in which the small dissociation of the acid solution is regarded as negligible. The concentrations of the individual molecular complexes are:

$$c_0 = \frac{n_0}{n_0 + n_1}, \quad c_1 = \frac{n_1}{n_0 + n_1}, \quad c_0' = \frac{n_0'}{n_0'} = 1$$

For the precipitation of solid succinic acid,

$$\nu_0 = 0, \quad \nu_1 = -1, \quad \nu_0' = 1,$$

and, therefore, from the condition of equilibrium expressed by equation (6)

$$-\log c_1 = \log K,$$

and then from equation (7)

$$\Delta Q = -RT^2 \frac{\partial \log c_1}{\partial T}.$$

This equation was used by van't Hoff to calculate the heat of solution ΔQ from the solubility of succinic acid at 0° and at 8:5 °C. The corresponding numbers were 2.88 and 4.22 in an arbitrary unit. The result of this approximation is:

$$\frac{\partial \log c_1}{\partial T} = \frac{\log_e 4.22 - \log_e 2.88}{8.5} = 0.04494,$$

from which for $T = 273$:

$$\Delta Q = -1.98 \cdot 273^2 . 0.04494 = -6{,}600 \text{ cal.},$$

which is the heat given up to the surroundings in the precipitation of a molecule of succinic acid. Berthelot found, through direct measurement, 6,700 calories for the heat of solution.

The absorption of a gas also comes under this category, e. g. carbonic acid in a liquid, such as water, of relatively unnoticeable smaller vapor pressure at not too high a temperature. For this system,

$$n_0 H_2 0,\ n_1 CO_2 \mid n_0' CO_2.$$

The vaporization of a molecule CO_2 corresponds to the values

$$\nu_0 = 0, \quad \nu_1 = -1, \quad \nu_0' = 1.$$

The condition of equilibrium is therefore again:

$$-\log c_1 = \log K,$$

and at a fixed temperature and a fixed pressure the concentration c_1 of the gas in the solution is constant.

The change of the concentration with p and T is obtained through substitution in equation (7):

$$\frac{\partial \log c_1}{\partial p} = \frac{\Delta V}{RT}, \quad \frac{\partial \log c_1}{\partial T} = -\frac{\Delta Q}{RT^2}.$$

ΔV is the change in volume of the system that occurs in the isobaric-isothermal vaporization of a molecule of CO_2, and ΔQ is the quantity of heat absorbed in the process from outside. Now, since ΔV represents the approximate volume of a molecule of gaseous carbonic acid:

$$\Delta V \cong \frac{RT}{p},$$

and then

$$\frac{\partial \log c_1}{\partial p} = \frac{1}{p},$$

which upon integrating yields:

$$\log c_1 = \log p + \text{const.}, \qquad c_1 = C \cdot p.$$

Therefore, the concentration of the dissolved gas is proportional to the pressure of the free gas above the solution (law of Henry and Bunsen). The factor of proportionality C, which furnishes a measure of the solubility of the gas, depends upon the heat effect in the same manner as in the previous example.

A number of important relations are easily derived as by-products of those found above, e. g., the Nernst laws concerning the influence of solubility, the Arrhenius theory of isohydric solutions, etc. All such may be obtained through the application of the general condition of equilibrium defined by equation (6).

VII. Osmotic Pressure.

This application is somewhat different in that Planck derives the osmotic pressure between a dilute solution of a dissolved substance and a pure solution. The same general methods apply in this case. For this system,

$$n_0 m_0,\, n_1 m_1,\, n_2 m_2, \cdots \mid n_0' m_0 \,.$$

The condition of equilibrium is again expressed by equation (6) for a change of state in which the temperature and the pressure in each phase is maintained constant. However, for the two phases on either side of the membrane, the ordinary hydrostatic pressure p in the first phase may be different from the pressure p' in the second phase.

The proof that equation (6) is valid for this application is similar to the prior example, utilizing the principle of increase of entropy. In this more general case, the external work in a given change is represented by the sum $pdV + p'dV'$, where V and V' denote the volumes of the two individual phases, while before V denoted the total volume of all phases. Accordingly, we use the following equation instead of instead of equation (7), to express the dependence of the constant K in (6) upon the pressure:

$$\frac{\partial \log K}{\partial p} = -\frac{\Delta V}{RT}, \quad \frac{\partial \log K}{\partial p'} = -\frac{\Delta V'}{RT}. \qquad (11)$$

The following symbolic representations,

$$\nu_0 = -1, \quad \nu_1 = 0, \quad \nu_2 = 0, \quad \cdots, \quad \nu_0' = 1,$$

express the fact that a molecule of the solvent passes out of the solution through the membrane into the pure solvent. Then in accordance with equation (6),

$$-\log c_0 = \log K,$$

and

$$c_0 = 1 - \frac{n_1 + n_2 + \cdots}{n_0}, \quad \frac{n_1 + n_2 + \cdots}{n_0} = \log K.$$

Here K depends only upon T, p and p'. With a pure solvent were present on both sides of the membrane, $c_0 = 1$, and $p = p'$, and therefore,

$$(\log K)_{p=p'} = 0.$$

By subtraction of the last two equations and taking the partial derivative:

$$\frac{n_1 + n_2 + \cdots}{n_0} = \log K - (\log K)_{p=p'} = \frac{\partial \log K}{\partial p}(p - p'),$$

and utilizing equation (11):

$$\frac{n_1 + n_2 + \cdots}{n_0} = -(p - p') \cdot \frac{\Delta V}{RT}.$$

Here ΔV denotes the change in volume of the solution due to the loss of a molecule of the solvent ($\nu_0 = -1$). The approximate volume of the whole solution is then

$$-\Delta V \cdot n_0 = V,$$

and

$$\frac{n_1 + n_2 + \cdots}{n_0} = (p - p') \cdot \frac{\Delta V}{RT}.$$

this equation contains the law of osmotic pressure due to van't Hoff, wherein the difference $(p - p')$ is the osmotic pressure of the solution.

The above applications of the relationships between the mathematical definition of state space and the measurements of the real world demonstrate the effects of changes in state. The validity of Planck's state equation (6) (for equilibrium) is well established by applying the thermodynamic equation (7) to these seven applications and comparing the results to known laboratory measurements.

The state equation is based on special reversible isothermal cycles under conditions of equilibrium

The system of equations that Planck developed can be applied to complex systems of molecular aggregations, which include various types of molecules of different numbers and masses and in multiple aggregations as was seen in most of the above applications. The advantage of dealing with energies and their quantum changes, rather than cyclic processes, was thus established. However, cyclic processes must still be considered in the analysis, as will be seen in the subsequent lectures.

Planck's Columbia Lectures

Chapter 3

Introduction to the Third Lecture:

Planck's next step is to extend the thermodynamic definitions of reversible and irreversible processes and the concept of entropy to atomic processes. Much of what follows is based on the lifetime work of Ludwig Boltzmann in the field of thermodynamics.

Boltzmann contributed so much to the advancement of science that he ranks among the greatest scientists of all time. Unfortunately, he was not properly recognized for his efforts during his lifetime. It is a sad note in history that he was so distraught by the lack of acclaim for his works that he committed suicide. Planck, however, properly lauds his efforts.

With the material in this chapter, the elementary foundation of Planck's theory is complete. The information necessary to solve the "mystery of "–1" is at hand, although it is not yet easily recognized. So far, rather than using empiricism in his methods, he establishes his theoretical mathematics and then verifies its validity in describing physical processes. In this lecture he applies his description of reversibility and irreversibility to the physical processes of matter.

THIRD LECTURE

THE ATOMIC THEORY OF MATTER.

The problem with which we shall be occupied in the present lecture is that of a closer investigation of the atomic theory of matter. It is, however, not my intention to introduce this theory with nothing further, and to set it up as something apart and disconnected with other physical theories, but I intend above all to bring out the peculiar significance of the atomic theory as related to the present general system of theoretical physics; for in this way only will it be possible to regard the whole system as one containing within itself the essential compact unity, and thereby to realize the principal object of these lectures.

Consequently it is self evident that we must rely on that sort of treatment which we have recognized in last week's lecture as fundamental. That is, the division of all physical processes into reversible and irreversible processes. Furthermore, we shall be convinced that the accomplishment of this division is only possible through the atomic theory of matter, or, in other words, that irreversibility leads of necessity to atomistics.

I have already referred at the close of the first lecture to the fact that in pure thermodynamics, which knows nothing of an atomic structure and which regards all substances as absolutely continuous, the difference between reversible and irreversible processes can only be defined in one way, which a priori carries a provisional character and does not withstand penetrating analysis. This appears immediately evident when one reflects that the purely thermodynamic definition of irreversibility which proceeds from the impossibility of the realization of certain changes in nature, as, e. g., the transformation of heat into work without compensation, has at the outset assumed a definite limit to man's mental capacity, while, however, such a limit is not indicated in reality. On the contrary: mankind is making every endeavor to press beyond the present boundaries of its capacity, and we hope that later on many things will be attained which, perhaps, many regard at present as impossible of accomplishment. Can it not happen then that a process, which up to the present has been regarded as irreversible, may be proved, through a new discovery or invention, to be reversible? In this case the whole structure of the second law would undeniably collapse, for the irreversibility of a single process conditions that of all the others.

It is evident then that the only means to assure to the second law real meaning consists in this, that the idea of irreversibility be made independent of any relationship to man and especially of all technical relations.

Now the idea of irreversibility harks back to the idea of entropy; for a process is irreversible when it is connected with an increase of entropy. The problem is hereby referred back to a proper improvement of the definition of entropy. In accordance with the original definition of Clausius, the entropy is measured by means of a certain reversible process, and the weakness of this definition rests upon the fact that many such reversible processes, strictly speaking all, are not capable of being carried out in practice. With some reason it may be objected that we have here to do, not with an actual process and an actual physicist, but only with ideal processes, so-called thought experiments, and with an ideal physicist who operates with all the experimental methods with absolute accuracy. But at this point the difficulty is encountered: How far do the physicist's ideal measurements of this sort suffice? It may be understood, by passing to the limit, that a gas is compressed by a pressure which is equal to the pressure of the gas, and is heated by a heat reservoir which possesses the same temperature as the gas, but, for example, that a saturated vapor shall be transformed through isothermal compression in a reversible manner to a liquid without at any time a part of the vapor being condensed, as in certain thermodynamic considerations is supposed, must certainly appear doubtful. Still more striking, however, is the liberty as regards thought experiments, which in physical chemistry is granted the theorist. With his semi-permeable membranes, which in reality are only realizable under certain special conditions and then only with a certain approximation, he separates in a reversible manner, not only all possible varieties of molecules, whether or not they are in stable or unstable conditions, but he also separates the oppositely charged ions from one another and from the undissociated molecules, and he is disturbed, neither by the enormous electrostatic forces which resist such a separation, nor by the circumstance that in reality, from the beginning of the separation, the molecules become in part dissociated while the ions in part again combine. But such ideal processes are necessary throughout in order to make possible the comparison ,of the entropy of the undissociated molecules with the entropy of the dissociated molecules; for the law of thermodynamic equilibrium does not permit in general of derivation in any other way, in case one wishes to retain pure thermodynamics as a basis. It must be considered remarkable that all these ingenious thought processes have so well found confirmation of their results in experience, as is shown by the examples considered by us in the last lecture.

If now, on the other hand, one reflects that in all these results every reference to the possibility of actually carrying out each ideal process has disappeared---there are certainly left relations between directly measurable quantities only, such as temperature, heat effect, concentration, etc.---the presumption forces itself upon one that perhaps the introduction as above of such ideal processes is at bottom a round-about method, and that the peculiar import of the principle of increase of entropy with all its consequences can be evolved from the original idea of irreversibility or, just as well, from the impossibility of perpetual motion of the second kind, just as the principle of conservation of energy has been evolved from the law of impossibility of perpetual motion of the first kind.

This step: to have completed the emancipation of the entropy idea from the experimental art of man and the elevation of the second law thereby to a real principle, was the scientific life's work of Ludwig Boltzmann. Briefly stated, it consisted in general of referring back the idea of entropy to the idea of probability. Thereby is also explained, at the same time, the significance of the above (p. 17) auxiliary term used by me; "preference" of nature for a definite state. Nature prefers the more probable states to the less probable, because in nature processes take place in the direction of greater probability. Heat goes from a body at higher temperature to a body at lower temperature because the state of equal temperature distribution is more probable than a state of unequal temperature distribution.

Through this conception the second law of thermodynamics is removed at one stroke from its isolated position, the mystery concerning the preference of nature vanishes, and the entropy principle reduces to a well understood law of the calculus of probability.

The enormous fruitfulness of so "objective" a definition of entropy for all domains of physics I shall seek to demonstrate in the following lectures. But today we have principally to do with the proof of its admissibility; for on closer consideration we shall immediately perceive that the new conception of entropy at once introduces a great number of questions, new requirements and difficult problems. The first requirement is the introduction of the atomic hypothesis into the system of physics. For, if one wishes to speak of the probability of a physical state, i. e., if he wishes to introduce the probability for a given state as a definite quantity into the calculation, this can only be brought about, as in cases of all probability calculations, by referring the state back to a variety of possibilities; i. e., by considering a finite number of a priori equally likely configurations (complexions) through each of which the state considered may be realized. The greater the number of complexions, the greater is the probability of the state. Thus, e. g., the probability of throwing a total of four with two ordinary six-sided dice is found through counting the complexions by which the throw with a total of four may be realized. Of these there are three complexions:

with the first die, 1, with the second die, 3, with the first die, 2, with the second die, 2, with the first die, 3, with the second die, 1.

On the other hand, the throw of two is only realized through a single complexion. Therefore, the probability of throwing a total of four is three times as great as the probability of throwing a total of two.

Now, in connection with the physical state under consideration, in order to be able to differentiate completely from one another the complexions realizing it, and to associate it with a definite reckonable number, there is obviously no other means than to regard it as made up of numerous discrete homogeneous elements ---for in perfectly continuous systems there exist no reckonable elements---and hereby the atomistic view is made a fundamental requirement. We have, therefore, to regard all bodies in nature, in so far as they possess an entropy, as constituted of atoms, and we therefore arrive in physics at the same conception of matter as that which obtained in chemistry for so long previously.

But we can immediately go a step further yet. The conclusions reached hold, not only for thermodynamics of material bodies, but also possess complete validity for the processes of heat radiation, which are thus referred back to the second law of thermodynamics. That radiant heat also possesses an entropy follows from the fact that a body which emits radiation into a surrounding diathermanous medium experiences a loss of heat and, therefore, a decrease of entropy. Since the total entropy of a physical system can only increase, it follows that one part of the entropy of the whole system, consisting of the body and the diathermanous medium, must be contained in the radiated heat. If the entropy of the radiant heat is to be referred back to the notion of probability, we are forced, in a similar way as above, to the conclusion that for radiant heat the atomic conception possesses a definite meaning. But, since radiant heat is not directly connected with matter, it follows that this atomistic conception relates, not to matter, but only to energy, and hence, that in heat radiation certain energy elements play an essential role. Even though this conclusion appears so singular and even though in many circles today vigorous objection is strongly urged against it, in the long run physical research will not be able to withhold its sanction from it, and the less, since it is confirmed by experience in quite a satisfactory manner. We shall return to this point in the lectures on heat radiation. I desire here only to mention that the novelty involved by the introduction of atomistic conceptions into the theory of heat radiation is by no means so revolutionary as, perhaps, might. Appear at the first glance. For there is, in my opinion at least, nothing which makes necessary the consideration of the heat processes in a complete vacuum as atomic, and it suffices to seek the atomistic features at the source of radiation, i. e., in those processes which have their play in the centres of emission and absorption of radiation. Then the Maxwellian electrodynamic differential equations can retain completely their validity for the vacuum, and, besides, the discrete elements of heat radiation are relegated exclusively to a domain which is still very mysterious and where there is still present plenty of room for all sorts of hypotheses.

Returning to more general considerations, the most important question comes up as to whether, with the introduction of atomistic conceptions and with the reference of entropy to probability, the content of the principle of increase of entropy is exhaustively comprehended, or whether still further physical hypotheses are required in order to secure the full import of that principle. If this important question had been settled at the time of the introduction of the atomic theory into thermodynamics, then the atomistic views would surely have been spared a large number of conceivable misunderstandings and justifiable attacks. For it turns out, in fact-and our further considerations will confirm this conclusion---that there has as yet nothing been done with atomistics which in itself requires much more than an essential generalization, in order to guarantee the validity of the second law.

We must first reflect that, in accordance with the central idea laid down in the first lecture (p. 7), the second law must possess validity as an objective physical law, independently of the individuality of the physicist. There is nothing to hinder us from imagining a physicist---we shall designate him a "microscopic" observer---whose senses are so sharpened that he is able to recognize each individual atom and to follow it in its motion. For this observer each atom moves exactly in accordance with the elementary laws which general dynamics lays down for it, and these laws allow, so far as we know, of an inverse performance of every process. Accordingly, here again the question is neither one of probability nor of entropy and its increase. Let us imagine, on the other hand, another observer, designated a "macroscopic" observer, who regards an ensemble of atoms as a homogeneous gas, say, and consequently applies the laws of thermodynamics to the mechanical and thermal processes within it. Then, for such an observer, in accordance with the second law, the process in general is an irreversible process. Would not now the first observer be justified in saying: "The reference of the entropy to probability has its origin in the fact that irreversible processes ought to be explained through reversible processes. At any rate, this procedure appears to me in the highest degree dubious. In any case, I declare each change of state which takes place in the ensemble of atoms designated a gas, as reversible, in opposition to the macroscopic observer." There is not the slightest thing, so far as I know, that one can urge against the validity of these statements. But do we not thereby place ourselves in the painful position of the judge who declared in a trial the correctness of the position of each separately of two contending parties and then, when a third contends that only one of the parties could emerge from the process victorious, was obliged to declare him also correct? Fortunately we find ourselves in a more favorable position. We can certainly mediate between the two parties without its being necessary for one or the other to give up his principal point of view. For closer consideration shows that the whole controversy rests upon a misunderstanding---a new proof of how necessary it is before one begins a controversy to come to an understanding with his opponent concerning the subject of the quarrel. Certainly, a given change of state cannot be both reversible and irreversible. But the one observer connects a wholly different idea with the phrase "change of state" than the other. What is then, in general, a change of state? The state of a physical system cannot well be otherwise defined than as the aggregate of all those physical quantities, through whose instantaneous values the time changes of the quantities, with given boundary conditions, are uniquely determined. If we inquire now, in accordance with the import of this definition, of the two observers as to what they understand by the state of the collection of atoms or the gas considered, they will give quite different answers. The microscopic observer will mention those quantities which determine the position and the velocities of all the individual atoms. There are present in the simplest case, namely, that in which the atoms may be considered as material points, six times as many quantities as atoms, namely, for each atom the three coordinates and the three velocity components, and in the case of combined molecules, still more quantities. For him the state and the progress of a process is then first determined when all these various quantities are individually given. We shall designate the state defined in this way the "micro-state." The macroscopic observer, on the other hand, requires fewer data. He will say that the state of the homogeneous gas considered by him is determined by the density, the visible velocity and the temperature at each point of the gas, and he will expect that, when these quantities are given, their time variations and, therefore, the progress of the process, to be completely determined in accordance with the two laws. Of thermodynamics, and therefore accompanied by an increase in entropy. In this connection he can call upon all the experience at his disposal, which will fully confirm his expectation. If we call this state the "macro-state," it is clear that the two laws: "the micro-changes of state are reversible" and "the macro-changes of state are irreversible," lie in wholly different domains and, at any rate, are not contradictory.

But now how can we succeed in bringing the two observers to an understanding? This is a question whose answer is obviously of fundamental significance for the atomic theory. First of all, it is easy to see that the macro-observer reckons only with mean values; for what he calls density, visible velocity and temperature of the gas are, for the micro-observer, certain mean values, statistical data, which are derived from the space distribution and from the velocities of the atoms in an appropriate manner. But the micro-observer cannot operate with these mean values alone, for, if these are given at one instant of time, the progress of the process is not determined throughout; on the contrary: he can easily find with given mean values an enormously large number of individual values for the positions and the velocities of the atoms, all of which correspond with the same mean values and which, in spite of this, lead to quite different processes with regard to the mean values. It follows from this of necessity that the micro-observer must either give up the attempt to understand the unique progress, in accordance with experience, of the macroscopic changes of state---and this would be the end of the atomic theory---or that he, through the introduction of a special physical hypothesis, restrict in a suitable manner the manifold of micro-states considered by him. There is certainly nothing to prevent him from assuming that not all conceivable micro-states are realizable in nature, and that certain of them are in fact thinkable, but never actually realized. In the formularization of such a hypothesis, there is of course no point of departure to be found from the principles of dynamics alone; for pure dynamics leaves this case undetermined. But on just this account any dynamical hypothesis, which involves nothing further than a closer specification of the micro-states realized in nature, is certainly permissible. Which hypothesis is to be given the preference can only be decided through comparison of the results to which the different possible hypotheses lead in the course of experience.

In order to limit the investigation in this way, we must obviously fix our attention only upon all imaginable configurations and velocities of the individual atoms which are compatible with determinate values of the density, the velocity and the temperature of the gas, or in other words: we must consider all the micro-states which belong to a determinate macro-state, and must investigate the various kinds of processes which follow in accordance with the fixed laws of dynamics from the different micro-states. Now, precise calculation has in every case always led to the important result that an enormously large number of these different micro-processes relate to one and the same macro-process, and that only proportionately few of the same, which are distinguished by quite special exceptional conditions concerning the positions and the velocities of neighboring atoms, furnish exceptions. Furthermore, it has also shown that one of the resulting macro-processes is that which the macroscopic observer recognizes, so that it is compatible with the second law of thermodynamics.

Here, manifestly, the bridge of understanding is supplied. The micro-observer needs only to assimilate in his theory the physical hypothesis that all those special cases in which special exceptional conditions exist among the neighboring configurations of interacting atoms do not occur in nature, or, in other words, that the micro-states are in elementary disorder. Then the uniqueness of the macroscopic process is assured and with it, also, the fulfillment of the principle of increase of entropy in all directions.

Therefore, it is not the atomic distribution, but rather the hypothesis of elementary disorder, which forms the real kernel of the principle of increase of entropy and, therefore, the preliminary condition for the existence of entropy. Without elementary disorder there is neither entropy nor irreversible process.[1] Therefore, a single atom can never possess an entropy; for we cannot speak of disorder in connection with it. But with a fairly large number of

---

1 To those physicists who, in spite of all this, regard the hypothesis of elementary disorder as gratuitous or as incorrect, I wish to refer the simple fact that in every calculation of a coefficient of friction, of diffusion, or of heat conduction, from molecular considerations, the notion of elementary disorder is employed, whether tacitly or otherwise, and that it is therefore essentially more correct to stipulate this condition instead of ignoring or concealing it. But he who regards the hypothesis of elementary disorder as self-evident, should be reminded that, in accordance with a law of H. Poincarre', the precise investigation concerning the foundation of which would here lead us too far, the assumption of this hypothesis for all times is unwarranted for a closed space with absolutely smooth walls,---an important conclusion, against which can only be urged the fact that absolutely smooth walls do not exist in nature.

atoms, say 100 or 1,000, the matter is quite different. Here, one can certainly speak of a disorder, in case that the values of the coordinates and the velocity components are distributed among the atoms in accordance with the laws of accident. Then it is possible to calculate the probability for a given state. But how is it with regard to the increase of entropy? May we assert that the motion of 100 atoms is irreversible? Certainly not; but this is only because the state of 100 atoms cannot be defined in a thermodynamic sense, since the process does not proceed in a unique manner from the standpoint of a macro-observer, and this requirement forms, as we have seen above, the foundation and preliminary condition for the definition of a thermodynamic state. If one therefore asks:

How many atoms are at least necessary in order that a process may be considered irreversible?, the answer is: so many atoms that one may form from them definite mean values which define the state in a macroscopic sense. One must reflect that to secure the validity of the principle of increase of entropy there must be added to the condition of elementary disorder still another, namely, that the number of the elements under consideration be sufficiently large to render possible the formation of definite mean values. The second law has a meaning for these mean values only; but for them, it is quite exact, just as exact as the law of the calculus of probability, that the mean value, so far as it may be defined, of a sufficiently large number of throws with a six-sided die, is 3 ½.

These considerations are, at the same time, capable of throwing light upon questions such as the following: Does the principle of increase of entropy possess a meaning for the so-called Brownian molecular movement of a suspended particle? Does the kinetic energy of this motion represent useful work or not? The entropy principle is just as little valid for a single suspended particle as for an atom, and therefore is not valid for a few of them, but only when there is so large a number that definite mean values can be formed. That one is able to see the particles and not the atoms makes no material difference; because the progress of a process does not depend upon the power of an observing instrument. The question with regard to useful work plays no role in this connection; strictly speaking, this possesses, in general, no objective physical meaning. For it does not admit of an answer without reference to the scheme of the physicist or technician who proposes to make use of the work in question. The second law, therefore, has fundamentally nothing to do with the idea of useful work (cf. First lecture, p. 15).

But, if the entropy principle is to hold, a further assumption is necessary, concerning the various disordered elements,--- an assumption which tacitly is commonly made and which we have not previously definitely expressed. It is, however, not less important than those referred to above. The elements must actually be of the same kind, or they must at least form a number of groups of like kind, e. g., constitute a mixture in which each kind of element occurs in large numbers. For only through the similarity of the elements does it come about that order and law can result in the larger from the smaller. If the molecules of a gas be all different from one another, the properties of a gas can never show so simple a law-abiding behavior as that which is indicated by thermodynamics. In fact, the calculation of the probability of a state presupposes that all complexions which correspond to the state are a priori equally likely. Without this condition one is just as little able to calculate the probability of a given state as, for instance, the probability of a given throw with dice whose sides are unequal in size. In summing up we may therefore say: the second law of thermodynamics in its objective physical conception, freed from anthropomorphism, relates to certain mean values which are formed from a large number of disordered elements of the same kind.

The validity of the principle of increase of entropy and of the irreversible progress of thermodynamic processes in nature is completely assured in this modularization. After the introduction of the hypothesis of elementary disorder, the microscopic observer can no longer confidently assert that each process considered by him in a collection of atoms is reversible; for the motion occurring in the reverse order will not always obey the requirements of that hypothesis. In fact, the motions of single atoms are always reversible, and thus far one may say, as before, that the irreversible processes appear reduced to a reversible process, but the phenomenon as a whole is nevertheless irreversible, because upon reversal the disorder of the numerous individual elementary processes would be eliminated. Irreversibility is inherent, not in the individual elementary processes themselves, but solely in their irregular constitution. It is this only which guarantees the unique change of the macroscopic mean values.

Thus, for example, the reverse progress of a frictional process is impossible, in that it would presuppose elementary arrangement of interacting neighboring molecules. For the collisions between any two molecules must thereby possess a certain distinguishing character, in that the velocities of two colliding molecules depend in a definite way upon the place at which they meet. In this way only can it happen that in collisions like directed velocities ensue and, therefore, visible motion.

Previously we have only referred to the principle of elementary disorder in its application to the atomic theory of matter. But it may also be assumed as valid, as I wish to indicate at this point, on quite the same grounds as those holding in the case of matter, for the theory of radiant heat. Let us consider, e. g., two bodies at different temperatures between which exchange of heat occurs through radiation. We can in this case also imagine a microscopic observer, as opposed to the ordinary macroscopic observer, who possesses insight into all the particulars of electromagnetic processes which are connected with emission and absorption, and the propagation of heat rays. The microscopic observer would declare the whole process reversible because all electrodynamic processes can also take place in the reverse direction, and the contradiction may here be referred back to a difference in definition of the state of a heat ray. Thus, while the macroscopic observer completely defines a monochromatic ray through direction, state of polarization, color, and intensity, the microscopic observer, in order to possess a complete knowledge of an electromagnetic state, necessarily requires the specification of all the numerous irregular variations of amplitude and phase to which the most homogeneous heat ray is actually subject. That such irregular variations actually exist follows immediately from the well known fact that two rays of the same color never interfere, except when they originate in the same source of light. But until these fluctuations are given in all particulars, the micro-observer can say nothing with regard to the progress of the process. He is also unable to specify whether the exchange of heat radiation between the two bodies leads to a decrease or to an increase of their difference in temperature. The principle of elementary disorder first furnishes the adequate criterion of the tendency of the radiation process, i.e., the warming of the colder body at the expense of the warmer, just as the same principle conditions the irreversibility of exchange of heat through conduction. However, in the two cases compared, there is indicated an essential difference in the kind of the disorder. While in heat conduction the disordered elements may be represented as associated with the various molecules, in heat radiation there are the numerous vibration periods, connected with a heat ray, among which the energy of radiation is irregularly distributed. In other words: the disorder among the molecules is a material one, while in heat radiation it is one of energy distribution. This is the most important difference between the two kinds of disorder; a common feature exists as regards the great number of uncoordinated elements required. Just as the entropy of a body is defined as a function of the macroscopic state, only when the body contains so many atoms that from them definite mean values may be formed, so the entropy principle only possesses a meaning with regard to a heat ray when the ray comprehends so many periodic vibrations, i.e., persists for so long a time, that a definite mean value for the intensity of the ray may be obtained from the successive irregular fluctuating amplitudes.

Now, after the principle of elementary disorder has been introduced and accepted by us as valid throughout nature, the fundamental question arises as to the calculation of the probability of a given state, and the actual derivation of the entropy therefrom. From the entropy all the laws of thermodynamic states of equilibrium, for material substances, and also for energy radiation, may be uniquely derived. With regard to the connection between entropy and probability, this is inferred very simply from the law that the probability of two independent configurations is represented by the product of the individual probabilities:

$$W = W_1 \cdot W_2,$$

while the entropy S is represented by the sum of the individual entropies:

$$S = S_1 + S_2.$$

Accordingly, the entropy is proportional to the logarithm of the probability:

$S = k \log W$. (12)

k is a universal constant. In particular, it is the same for atomic as for radiation configurations, for there is nothing to prevent us assuming that the configuration designated by 1 is atomic, while that designated by 2 is a radiation configuration. If k has been calculated, say with the aid of radiation measurements, then k must have the same value for atomic processes. Later we shall follow this procedure, in order to utilize the laws of heat radiation in the kinetic theory of gases. Now, there remains, as the last and most difficult part of the problem, the calculation of the probability W of a given physical configuration in a given macroscopic state. We shall treat today, by way of preparation for the quite general problem to follow, the simple problem: to specify the probability of a given state for a single moving material point, subject to given conservative forces. Since the state depends upon 6 variables: the 3 generalized coordinates $\varphi_1, \varphi_2, \varphi_3,$ and the three corresponding velocity components, $\dot{\varphi}_1, \dot{\varphi}_2, \dot{\varphi}_3$, and since all possible values of these 6 variables constitute a continuous manifold, the probability sought is, that these 6 quantities shall lie respectively within certain infinitely small intervals, or, if one thinks of these 6 quantities as the rectilinear orthogonal coordinates of a point in an ideal six-dimensional space, that this ideal "state point" shall fall within a given, infinitely small "state domain." Since the domain is infinitely small, the probability will be proportional to the magnitude of the domain and therefore proportional to

$$\int d\varphi_1 \cdot d\varphi_2 \cdot d\varphi_3 \cdot d\dot{\varphi}_1 \cdot d\dot{\varphi}_2 \cdot d\dot{\varphi}_3 .$$

But this expression cannot serve as an absolute measure of the probability, because in general it changes in magnitude with the time, if each state point moves in accordance with the laws of motion of material points, while the probability of a state which follows of necessity from another must be the same for the one as the other. Now, as is well known, another integral quite similarly formed, may, be specified in place of the one above, which possesses the special property of not changing in value with the time. It is only necessary to employ, in addition to the general coordinates φ_1, φ_2, φ_3, the three so-called momenta ψ_1, ψ_2, ψ_3, in place of the three velocities, $\dot{\varphi}_1, \dot{\varphi}_2, \dot{\varphi}_3$, as the determining coordinates of the state. These are defined in the following way:

$$\psi_1 = \left(\frac{\partial H}{\partial \dot{\varphi}_1}\right)_\varphi , \quad \psi_2 = \left(\frac{\partial H}{\partial \dot{\varphi}_2}\right)_\varphi , \quad \dot{\varphi}_3 = \left(\frac{\partial H}{\partial \dot{\varphi}_3}\right)_\varphi ,$$

wherein H denotes the kinetic potential (Helmholz). Then, in Hamiltonian form, the equations of motion are:

$$\psi_1 = \frac{d\psi_1}{dt} = -\left(\frac{\partial E}{\partial \varphi_1}\right)_\psi , \cdots, \quad \dot{\varphi}_1 = \frac{d\varphi_1}{dt} = -\left(\frac{\partial E}{\partial \psi_1}\right)_\varphi , \cdots ,$$

(E is the energy), and from these equations[2] follows the "condition of incompressibility":

$$\frac{\partial \dot{\varphi}_1}{\partial \varphi} + \frac{\partial \psi_1}{\partial \psi_1} + \cdots = 0$$

Referring to the six-dimensional space represented by the coordinates $\varphi_1, \varphi_2, \varphi_3, \psi_1, \psi_2, \psi_3$, this equation states that the magnitude of an arbitrarily chosen state domain, viz.:

$$\int d\varphi_1 \cdot d\varphi_2 \cdot d\varphi_3 \, d\psi_1 \cdot d\psi_2 \cdot d\psi_3$$

does not change with the time, when each point of the domain changes its position in accordance with the laws of motion of material points. Accordingly, it is made possible to take the magnitude of this domain as a direct measure for the probability that the state point falls within the domain.

From the last expression, which can be easily generalized for the case of an arbitrary number of variables, we shall calculate later the probability of a thermodynamic state, for the case of radiant energy as well as that for material substances.

2 There may have been a typographical error in these two equations. The following corrections are suggested:

$$\dot{\psi}_1 = \frac{d\psi_1}{dt} = -\left(\frac{\partial E}{\partial \varphi_1}\right)_{\psi}, \quad \frac{\partial \dot{\varphi}_1}{\partial \varphi} + \frac{\partial \dot{\psi}_1}{\partial \psi_1} + \cdots = 0.$$

Planck's Columbia Lectures

Summary of the Third Lecture

Planck began this lecture by defining the role of reversibility and irreversibility for molecular processes. A molecule can be split into other molecular forms or into ions. Molecules can also be re-formed by suitable processes, such as chemical separation methods that involve semi-permeable membranes under the proper conditions. However, not all of the molecules or ions undergo a reversible process. The electrostatic forces that resist molecular separation must be considered as part of this process. The reversible processes are dependent upon forces that are not directly measurable, and theory is therefore dependent upon "thought processes". The irreversible processes, however, allow the successful comparison of the entropy of undissociated molecules with the entropy of dissociated molecules (Lecture II).

The presumptive measurable variables necessary to define ideal processes are temperature, heat effect, concentration, etc., and the applicable principles are the impossibility of perpetual motion and energy loss. In resolving these issues, Planck referred to the efforts of Boltzmann, who related the idea of entropy to the idea of probability. Planck referred to the law of the preference of nature and added the assertion that nature prefers the more probable states to the less probable. Nature processes take place in the direction of greater probability. An example is that heat goes from a body that is at high temperature to one that is at low temperature. The entropy principle is thus reduced to the law of the calculus of probability.

Planck provided evidence of the admissibility of this principle by referring the probability of a physical state back to a variety of possible configurations through which the state may be realized. As an example, he chose the throw of a die with six equal numbered sides, and each side of the die is equally probable to occur in a throw. A throw of two dice that results in total count of two has only one possible "complexion", while the probability of throwing a total count of four is three times as great (the three complexions are 1 and 3, 2 and 2, and 3 and 1).

At this point, Planck made a number of assumptions. All bodies in nature must be constituted of atoms, in accordance with the concept of matter in chemistry. All physical atomic states must be differentiated from one another by a definite reckonable number. He thus introduced the initial concept of quantum theory: "...in perfectly continuous systems there exist no reckonable elements --- and hereby the atomistic view is made a fundamental requirement".[3]

The entropy principle is not valid for a single "suspended particle" (atom), because entropy is a function of disordered elements and probable states.

The difficulties of measuring the action of a single atom are avoided by measuring only mean values of a system consisting of large numbers of like atoms in disorder. The probability of a state presupposes that all complexions that correspond to a state are equally likely.

---

3 Modern classical analysis is not limited to just continuous systems. Step functions and jump functions can usually be handled rather easily using modern linear analysis. Nonlinear systems, which may or may not be continuous, are much more difficult to analyze, but classical analysis has the ability to produce exact solutions for certain systems of various complexity. The problem of converting from energy functions to force functions is discussed in Chapter 10. Quantum concepts are very compatible with classical analysis, notwithstanding contrary assertions of numerous physicists.

With the hypothesis of elementary disorder, it becomes clear that reversible processes are highly unlikely to occur for large numbers of atoms by virtue of the laws of probability. Each atom can be moving in any direction and at any velocity that is possible in the three dimensions of space. The particular path and velocity of the atom that would reverse this process is therefore highly unlikely. A single atom is reversible, while the assemblage of atoms is irreversible. The microscopic observer would be able to see the actions of the atoms, while the macroscopic observer see only the effects of the average values of the atoms.

He went further in this assertion, applying it also to heat radiation. His argument was that heat must also possess entropy since a body that emits radiation experiences a loss of heat and a decrease in entropy. This finite amount of heat loss has definite meaning for atomic concepts.[4] The microscopic observer would declare the atomic process to be reversible because all electrodynamic processes can take place in the reverse direction. The macroscopic observer would define a monochromatic ray through direction, polarization, color and intensity, while the microscopic observer notes all of the irregular variations of amplitude and phase of of the homogeneous heat ray. Heat rays exhibit these types of variations and are lacking frequency coherence.

Entropy is therefore related to disorder and probability, and the next step was to define the mathematical relationships. The probability of two independent configurations having probabilities W_1 and W_2 is

4 Einstein is given credit for this quantum concept in some physics texts, although it is somewhat different, as will be seen in the last lecture of this series.

$$W = W_1 \cdot W_2$$

and the total entropy is

$$S = S_1 + S_2 = k \log W,$$

which applies to atomic and radiation configurations.

The last step was to define the six general state coordinates. The three space coordinates are φ_1, φ_2, φ_3, and the probability of the state is proportional to the magnitude of the domain, which is a function of the space coordinates and velocities,

$$\int d\varphi_1 d\varphi_2 d\varphi_3 d\dot{\varphi}_1 d\dot{\varphi}_2 d\dot{\varphi}_3 .$$

The velocities in the above equation are a function of time, so it is not an absolute measure of the probability. This time variance is removed by substituting ψ_1, ψ_2, ψ_3, which are defined as the differential ratios of the kinetic potential to the change in state, for the three velocities in the above equation,

$$\int d\varphi_1 d\varphi_2 d\varphi_3 d\psi_1 d\psi_2 d\psi_3$$

which represents the magnitude of a given state domain and the probability that the state point falls within the domain.[5]

5 Note that this equation can be compared to the *matter wave function*, Ψ, of wave mechanics which quantifies a localized bounding particle. Max Born asserted that the probability that a particle will be near a point in space is Ψ^2, which has been used to

This equation will be used to calculate the probability of the thermodynamic states of both radiant energy and material substances.

Planck thus formed the foundation upon which his theory is based. The next step will be to perform the actual calculations involving the laws of probability.

---

find the probability of the separation between the nucleus and the electron of the hydrogen atom.

Chapter 4

Introduction to the Fourth Lecture

In this lecture, Planck develops the probability equations for the entropy of a material substance in the form of a monatomic gas and the concepts of state space. As in the second lecture, all sorts of good things are seen to come from the analysis. Applying the entropy equation and energy values, he is able to derive more of the laws of physics and relate the entire ensemble of equations to real world measurements.

In Chapter 2, the analysis was based on dilute solutions in equilibrium. In this chapter, the gas need not be in equilibrium, the density need not be uniform, and the velocities and coordinates of the molecules can be arbitrary and without uniformity.

The definitions of "microspace" and "macrospace" are further delineated, and they are utilized to simplify the very complex problem of dealing with huge numbers of molecules. The probability equations for the microstates of the microspace domains are used to develop the probability of a macrostate. The methods that Planck uses to relate entropy, probability of the macro-states, energy and volume of the system are very thorough. However, the resulting final manipulations are relatively simple and produce important results. Further clues to the secret of "–1" lie within these pages.

FOURTH LECTURE

THE EQUATION OF STATE FOR A MONATOMIC GAS.

My problem today is to utilize the general fundamental laws concerning the concept of irreversibility, which we established in the lecture of yesterday, in the solution of: a definite problem: the calculation of the entropy of an ideal monatomic gas in a given state, and the derivation of all its thermodynamic properties. The way in which we have to proceed is prescribed for us by the general definition of entropy:

$$S = k \log W. \qquad (13)$$

The chief part of our problem is the calculation of W for a given state of the gas, and in this connection there is first required a more precise investigation of that which is to be understood as the state of the gas. Obviously, the state is to be taken here solely in the sense of the conception which we have called macroscopic in the last lecture. Otherwise, a state would possess neither probability nor entropy. Furthermore, we are not allowed to assume a condition of equilibrium for the gas. For this is characterized through the further special condition that the entropy for it is a maximum. Thus, an unequal distribution of density may exist in the gas; also, there may be present an arbitrary number of different currents, and in general no kind of equality between the various velocities of the molecules is to be assumed. The velocities, as the coordinates of the molecules, are rather to be taken a priori as quite arbitrarily given, but in order that the state, considered in a macroscopic sense,may be assumed as known, certain mean values of the densities and the velocities must exist. Through these mean values the state from a macroscopic standpoint is completely characterized.

The conditions mentioned will all be fulfilled if we consider the state as given in such manner that the number of molecules in a sufficiently small macroscopic space, but which, however, contains a very large number of molecules, is given, and furthermore, that the (likewise great) number of these molecules is given, which are found in a certain macroscopically small velocity domain, i.e., whose velocities lie within certain small intervals. If we call the coordinates $x, y, z,$ and the velocity components $\dot{x}, \dot{y}, \dot{z},$ then this number will be proportional to[1]

$$dx \cdot dy \cdot dz \cdot d\dot{x} \cdot d\dot{y} \cdot d\dot{z} = \sigma \, .$$

It will depend, besides, upon a finite factor of proportionality which may be an arbitrarily given function $f(x, y, z, \dot{x}, \dot{y}, \dot{z})$ of the coordinates and the velocities, and which has only the one condition to fulfill that

$$\sum f \cdot \sigma = N, \qquad (14)$$

where N denotes the total number of molecules in the gas. We are now concerned with the calculation of the probability W of that state of the gas which corresponds to the arbitrarily given distribution function f.

---

1 We can call a a "macro-differential" in contradistinction to the micro-differentials which are infinitely small with reference to the dimensions of a molecule. I prefer this terminology for the discrimination between "physical" and "mathematical" differentials in spite of the inelegance of phrasing, because the macro-differential is also just as much mathematical as physical and the micro-differential just as much physical as mathematical.

The probability that a given molecule possesses such coordinates and such velocities that it lies within the domain σ is expressed, in accordance with the final result of the previous lecture, by the magnitude of the corresponding elementary domain:

$$d\varphi_1,\ d\varphi_2,\ d\varphi_3,\ d\psi_1,\ d\psi_2\ d\psi_3,$$

therefore, since here

$$\varphi_1 = x, \varphi_2 = y, \varphi_3 = z, \psi_1 = m\dot{x}, \psi_2 = m\dot{y}, \psi_3 = m\dot{z},$$

(m the mass of a molecule) by

$$m^3\sigma.$$

Now we divide the whole of the six dimensional "state domain" containing all the molecules into suitable equal elementary domains of the magnitude $m^3\sigma$. Then the probability that a given molecule falls in a given elementary domain is equally great for all such domains. Let P denote the number of these equal elementary domains. Next, let us imagine as many dice as there are molecules present, i.e., N, and each die to be provided with P equal sides. Upon these P sides we imagine numbers $1, 2, 3, \cdots$ to P, so that each of the P sides indicates a given elementary domain. Then each throw with the N dice corresponds to a given state of the gas, while the number of dice which show a given number corresponds to the molecules which lie in the elementary domain considered. In accordance with this, each single die can indicate with the same probability each of the numbers from 1 to P, corresponding to the circumstance that each molecule may fall with equal probability in any one of the P elementary domains. The probability W sought, of the given state of the molecules, corresponds, therefore, to the number of different kinds of throws (complexions) through which is realized the given distribution f. Let us take, e.g., N equal to 10 molecules (dice) and $P = 6$ elementary domains (sides) and let us imagine the state so given that there are

3 molecules in 1st elementary domain
4 molecules in 2d elementary domain
0 molecules in 3d elementary domain
1 molecule in 4th elementary domain
0 molecules in 5th elementary domain
2 molecules in 6th elementary domain,

then this state, e.g., may be realized through a throw for which the 10 dice indicate the following numbers:

1st	2d	3d	4th	5th	6th	7th	8th	9th	10th	
2	6	2	1	1	2	6	2	1	4.	(15)

Under each of the characters representing the ten dice stands the number which the die indicates in the throw. In fact,

3 dice show the figure 1
4 dice show the figure 2
0 dice show the figure 3
1 die shows the figure 4
0 dice show the figure 5
2 dice show the figure 6.

The state in question may likewise be realized through many other complexions of this kind. The number sought of all possible complexions is now found through consideration of the number series indicated in (15). For, since the number of molecules (dice) is given, the number series contains a fixed number of elements ($10 = N$). Furthermore, since the number of molecules falling in an elementary domain is given, each number, in all permissible complexions, appears equally often in the series. Finally, each change of the number configuration conditions a new complexion. The number of possible complexions or the probability W of the given state is therefore equal to the number of possible permutations with repetition under the conditions mentioned. In the simple example chosen, in accordance with a well known formula, the probability is

$$\frac{10!}{3!4!0!1!0!2!} = 12,600$$

Therefore, in the general case:

$$W = \frac{N!}{\prod (f \cdot \sigma)!}.$$

The sign $\prod$ denotes the product extended over all of the P elementary domains.

From this there results, in accordance with equation (13), for the entropy of the gas in the given state:

$$S = k \log N! - k \Sigma \log (f \bullet \sigma)!.$$

The summation is to be extended over all domains σ. Since $f \cdot \sigma$ is a large quantity, Stirling's formula may be employed for its factorial, which for a large number n is expressed by:

$$n! = \left(\frac{n}{e}\right)^n \sqrt{2\pi\, n}\,, \qquad (16)$$

therefore, neglecting unimportant terms:

$$\log n! \;=\; n\,(\log n \,-\, 1)\,;$$

and hence:

$$S = k \log N\,! \;-\; k\,\Sigma\, f\sigma\,(\log\,[f \cdot \sigma\,] \;-\; 1),$$

or, if we note that σ and $N = \Sigma\, f\sigma$ remain constant in all changes of state:

$$S \;=\; \text{const} - k\,\Sigma\, f \cdot \log f \cdot \sigma\,. \qquad (17)$$

This quantity is, to the universal factor $(-k)$, the same as that which L. Boltzmann denoted by H, and which he showed to vary in one direction only for all changes of state.

In particular, we will now determine the entropy of a gas in a state of equilibrium, and inquire first as to that form of the law of distribution which corresponds to thermodynamic equilibrium. In accordance with the second law of thermodynamics, a state of equilibrium is characterized by the condition that with given values of the total volume V and the total energy E, the entropy S assumes its maximum value. If we assume the total volume of the gas

$$V = \int dx \cdot dy \cdot dz\,,$$

and the total energy

$$E = \frac{m}{2} \sum (\dot{x}^2 + \dot{y}^2 + \dot{z}^2) f \sigma \qquad (18)$$

as given, then the condition:

$$\delta S = 0$$

must hold for the state of equilibrium, or, in accordance with (17):

$$\Sigma(\log f + 1) \cdot \delta f \cdot \sigma = 0, \qquad (19)$$

wherein the variation δf refers to an arbitrary change in the law of distribution, compatible with the given values of N, V and E.

Now we have, on account of the constancy of the total number of molecules N, in accordance with (14):

$$\Sigma \, \delta f \cdot \sigma = 0.$$

and, on account of the constancy of the total energy, in accordance with (18):

$$\sum (\dot{x}^2 + \dot{y}^2 + \dot{z}^2) \cdot \delta f \cdot \sigma = 0.$$

Consequently, for the fulfillment of condition (19) for all permissible values of δf, it is sufficient and necessary that

$$\sum (\dot{x}^2 + \dot{y}^2 + \dot{z}^2) \cdot \delta f \cdot \sigma = 0.$$

$$\log f + \beta (\dot{x}^2 + \dot{y}^2 + \dot{z}^2) = \text{const},$$

or:

$$f = \alpha e^{-\beta(\dot{x}^2 + \dot{y}^2 + \dot{z}^2)},$$

wherein α and β are constants. In the state of equilibrium, therefore, the space distribution of molecules is uniform, i.e., independent of x, y, z, and the distribution of velocities is the well known Maxwellian distribution.

The values of the constants α and β are to be found from those of N, V and E. For the substitution of the value found for f in (14) leads to:

$$N = V\alpha\,(\frac{\pi}{\beta})^{\frac{3}{2}},$$

and the substitution of f in (18) leads to:

$$E = \frac{3}{4} V m (\frac{\pi}{\beta})^{\frac{3}{2}}.$$

From these equations it follows that:

$$\alpha = \frac{N}{V} \cdot (\frac{3mN}{4\pi E})^{\frac{3}{2}}, \quad \beta = \frac{3mN}{4E},$$

and hence finally, in accordance with (17), the expression for the entropy S of the gas in a state of equilibrium with given values for N, V and E is:

$$S = \text{const} + kN(\tfrac{3}{2}\log E + \log V). \qquad (20)$$

The additive constant contains terms in N and m, but not in E and V.

The determination of the entropy here carried out permits now the specification directly of the complete thermodynamic behavior of the gas, viz., of the equation of state, and of the values of the specific heats. From the general thermodynamic definition of entropy:

$$dS = \frac{dE + pdV}{T}$$

are obtained the partial differential quotients of S with regard to E and V respectively:

$$\left(\frac{dS}{\partial E}\right)_V = \frac{1}{T}, \quad \left(\frac{dS}{\partial V}\right)_E = \frac{p}{T}.$$

Consequently, with the aid of (20):

$$\left(\frac{dS}{\partial E}\right)_V = \frac{3}{2}\frac{kN}{E} = \frac{1}{T}, \qquad (21)$$

and

$$\left(\frac{dS}{\partial V}\right)_E = \frac{kN}{V} = \frac{p}{T}, \qquad (22)$$

The second of these equations:

$$p = \frac{kNT}{V}$$

contains the laws of Boyle, Gay Lussac and Avogadro, the latter because the pressure depends only upon the number N, and not upon the constitution of the molecules. Writing it in the ordinary form:

$$p = \frac{RnT}{V},$$

where n denotes the number of gram molecules or mols of the gas, referred to O_2 = 32g, and R the absolute gas constant:

$$\mathrm{R} = 8.315 \cdot 10^7 \ \text{erg/deg},$$

we obtain by comparison:

$$k = \frac{Rn}{N}. \quad (23)$$

and hence:

$$k = \omega R. \qquad (24)$$

From this, if ω is given, we can calculate the universal constant k, and conversely.

The equation (21) gives:

$$E = \tfrac{3}{2} k N T. \qquad (25)$$

Now since the energy of an ideal gas is given by:

$$E = A n c_v T,$$

wherein c_v denotes in calories the heat capacity at

constant volume of a mol, A the mechanical equivalent of heat:

$$A = 4.19 \cdot 10^7 \text{ erg/cal},$$

it follows that:

$$c_v = \frac{3kN}{2An},$$

and, having regard to (23), we obtain:

$$c_v = \frac{3}{2}\frac{R}{A} = 3.0, \qquad (26)$$

the mol heat in calories of any monatomic gas at constant volume.

For the mol heat c_p at constant pressure we have from the first law of thermodynamics

$$c_p - c_v = \frac{R}{A},$$

and, therefore, having regard to (26):

$$c_p = 5, \quad \frac{c_p}{c_v} = \frac{5}{3},$$

a known result for monatomic gases.

The mean kinetic energy L of a molecule is obtained from (25):

$$L = \frac{E}{N} = \tfrac{3}{2} k T. \qquad (27)$$

You notice that we have derived all these relations through the. identification of the mechanical with the thermodynamic expression for the entropy, and from this you recognize the fruitfulness of the method here proposed.

But a method can first demonstrate fully its usefulness when we utilize it, not only to derive laws which are already known, but when we apply it in domains for whose investigation there at present exist no other methods. In this connection its application affords various possibilities. Take the case of a monatomic gas which is not sufficiently attenuated to have the properties of the ideal state; there are here, as pointed out by J. D. van der Waals, two things to consider: (1) the finite size of the atoms, (2) the forces which act among the atoms. Taking account of these involves a change in the value of the probability and in the energy of the gas as well, and, so far as can now be shown, the corresponding change in the conditions for thermodynamic equilibrium leads to an equation of state which agrees with that of van der Waals. Certainly there is here a rich field for further investigations, of greater promise when experimental tests of the equation of state exist in larger number.

Another important application of the theory has to do with heat radiation, with which we shall be occupied the coming week. We shall proceed then in a similar way as here, and shall be able from the expression for the entropy of radiation to derive the thermodynamic properties of radiant heat.

Today we will refer briefly to the treatment of

polyatomic gases. I have previously, upon good grounds, limited the treatment to monatomic molecules; for up to the present real difficulties appear to stand in the way of a generalization, from the principles employed by us, to include polyatomic molecules; in fact, if we wish to be quite frank, we must say that a satisfactory mechanical theory of polyatomic gases has not yet been found. Consequently, at present we do not know to what place in the system of theoretical physics to assign the processes within a molecule---the intra-molecular processes. We are obviously confronted by puzzling problems. A noteworthy and much discussed beginning was, it is true, made by Boltzmann, who introduced the most plausible assumption that for intra-molecular processes simple laws of the same kind hold as for the motion of the molecules themselves, i. e., the general equations of dynamics. It is easy then, in fact, to proceed to the proof that for a monatomic gas the molecular heat c_v must be greater than 3 and that consequently, since the difference c_p - c_v is always equal to 2, the ratio is

$$\frac{c_p}{c_v} = \frac{c_v + 2}{c_v} < \frac{5}{3}.$$

This conclusion is completely confirmed by experience. But this in itself does not confirm the assumption of Boltzmann; for, indeed, the same conclusion is reached very simply from the assumption that there exists intra-molecular energy which increases with the temperature. For then the molecular heat of a polyatomic gas must be greater by a corresponding amount than that of a monatomic gas.

Nevertheless, up to this point the Boltzmann theory never leads to contradiction with experience. But so soon as one seeks to draw special conclusions concerning the magnitude of the specific heats hazardous difficulties arise; I will refer to only one of them. If one assumes the Hamiltonian equations of mechanics as applicable to intra-molecular motions, he arrives of necessity at the law of "uniform distribution of energy," which asserts that under certain conditions, not essential to consider here, in a thermodynamic state of equilibrium the total energy of the gas is distributed uniformly among all the individual energy phases corresponding to the independent variables of state, or, as one may briefly say; the same amount of energy is associated with every independent variable of state. Accordingly, the mean energy of motion of the molecules $\frac{1}{2} kT$, corresponding to a given direction in space, is the same as for any other direction, and, moreover, the same for all the different kinds of molecules, and ions; also for all suspended particles (dust) in the gas, of whatever size, and, furthermore, the same for all kinds of motions of the constituents of a molecule relative to its centroid. If one now reflects that a molecule commonly contains, so far as we know, quite a large number of different freely moving constituents, certainly, that a normal molecule of a monatomic gas, e.g., mercury, possesses numerous freely moving electrons, then, in accordance with the law of uniform energy distribution, the intra-molecular energy must constitute a much larger fraction of the whole specific heat of the gas, and therefore c_p/c_v must turn out much smaller, than is consistent with the measured values. Thus, e.g., for an atom of mercury, in accordance with the measured value of $c_p/c_v = 5/3$, no part whatever of the heat added may be assigned to the intra-molecular energy. Boltzmann and others, in order to eliminate this contradiction, have fixed upon the possibility that, within the time of observation of the specific heats, the vibrations of the constituents (of a molecule) do not change appreciably with respect to one another, and come later with their progressive motion so slowly into heat equilibrium that this process is no longer capable of detection through observation. Up to now no such delay in the establishment of a state of equilibrium has been observed. Perhaps it would be productive of results if in delicate measurements special attention were paid the question as to whether observations which take a longer time lead to a greater value of the mol-heat, or, what comes to the same thing, a smaller value of c_p/c_v, than observations lasting a shorter time.

If one has been made mistrustful through these considerations concerning the applicability of the law of uniform energy distribution to intra-molecular processes, the mistrust is accentuated upon the inclusion of the laws of heat radiation. I shall make mention of this in a later lecture.

When we pass from stable atoms to the unstable atoms of radioactive substances, the principles following from the kinetic gas theory lose their validity completely. For the striking failure of all attempts to find any influence of temperature upon radioactive phenomena, shows us that an application here of the law of uniform energy distribution is certainly not warranted. It will, therefore, be safest meanwhile to offer no definite conjectures with regard to the nature and the laws of these noteworthy phenomena, and to leave this field for further development to experimental research alone, which, I may say, with every day throws new light upon the subject.

Planck's Columbia Lectures

Summary of the Fourth Lecture

Planck's definitions of macrostates, microstates, elementary domains and macro-differentials and how he uses these definitions to determine probabilities of given states are very important.

The definition of entropy is dependent upon the probability of a given state of the gas, and the first half of this lecture was devoted to developing the applicable probability equations. Each molecule in the gas is associated with a probable energy state. Every macrostate is associated with a small space in which a large number of molecules have velocities that lie within certain small intervals. The number of the molecules in this macrospace is proportional to

$$dx \cdot dy \cdot dz \cdot d\dot{x} \cdot d\dot{y} \cdot d\dot{z} = \sigma ,$$

where σ is termed a "macro-differential" that is proportional to the number of molecules in the macrospace.

The probability factor is dependent upon an arbitrary function and the total number of molecules in the gas,

$$\sum f \cdot \sigma = N, \quad (14)$$

and the proportionality factor, $f(dx \cdot dy \cdot dz \cdot d\dot{x} \cdot d\dot{y} \cdot d\dot{z})$, is a function of the above variables.

The magnitude of an elementary domain is

$$d\varphi_1 \cdot d\varphi_2 \cdot d\varphi_3 \cdot d\psi_1 \cdot d\psi_2 \cdot d\psi_3 ,$$

wherein the states are defined as

$$\varphi_1 = x,\ \varphi_1 = y,\ \varphi_1 = z,\ \psi_1 = m\dot{x},\ \psi_2 = \dot{y},\ \psi_3 = \dot{z}, m = \text{mass}.$$

The coordinates and velocities for molecules that lie within each elementary domain σ are within narrow limits, which is fundamental to the calculation of the probability that a molecule lies within this domain. The state domain of a molecule was thus defined in terms of six dimensions; the three dimensions of space and the three dimensional velocities associated with the molecule in this space. The gas is divided into equally probable elementary domains of magnitude $m^3\sigma$.

Planck described an example of N dice that have P uniquely numbered sides that produce probabilities of a throw. The probability, W, of a given state of the molecules corresponds to the number of different kinds of throws of the dice by which it can be realized. In this case, $N = 10$ molecules (dice), and $P = 6$ elementary domains (sides). The unique state chosen is for this example is 3 molecules in the first domain, 4 in the second, 0 in the third, 1 in the fourth, 0 in the fifth, and 2 in the sixth, which can be realized in one instance by a throw resulting in the following numbers on the dice that correspond to each of the six elementary domains:

2, 6, 2, 1, 1, 2, 6, 2, 1 4.

There are 3 dice with the number 1, 4 dice with the number 2, which corresponds to 4 molecules in the second state, 2 dice with the number 6 relates to 2 molecules in the sixth state, etc..

The above throw is but one way to obtain this particular combination of numbers (or molecular state). The total number of combinations by which this set of numbers can be obtained is

$$\frac{10!}{3!4!0!1!0!2!} = 12,600.$$

The primes (!) in the above equation are defined as the factorials $n(n\text{-}1)(n\text{-}2)\text{---}1$. Planck called the number of allowable combinations the "probability", whereas in today's terminology it is the <u>inverse of this number</u>. There are 10! possible combinations of ten dice with six sides, and there is one chance in 12,600 of getting the above combination of numbers (states) in a single throw. If there were but a single state that predominates for all of the dice, then all ten dice would have the same number for all throws, resulting in only one single combination that is possible. The possibility of getting 10 ones is therefore one in 10!, and the entropy
($S = k \log W$) is obtained from the maximum number of possibilities ($W = 3{,}628{,}800$ in this example).

The next steps involve a few mathematical manipulations. The probability function can be expressed in general terms,

$$W = \frac{N!}{\prod (f \cdot \sigma)!},$$

where the symbol $\prod$ represents a product series of the terms within the primed parentheses that extend over the P elementary domains.

Substituting into the entropy equation and taking logs,

$$S = k \log N! - k \sum \log(f \cdot \sigma)!$$

for which the summation is over all of the σ elementary domains.

The factorials are difficult to manipulate, but when ($f \cdot \sigma$) is a large number Stirling's formula can be employed., resulting in the approximation

$$n! = \left(\frac{n}{e}\right)^n \sqrt{2\pi n}. \quad (16)$$

A further simplification is obtained by another approximation

$$\log n! = n(\log n - 1),$$

and substituting,

$$S = k \log N! - k \sum f\sigma \left(\log[f \cdot \sigma] - 1\right).$$

Finally, since σ and $N = \sum f\sigma$ are constant for all changes in state,

$$S = \text{const} - k \sum f \cdot \log f \cdot \sigma. \quad (17)$$

Planck thus arrived at a comparatively simple equation for the entropy of gas in terms of the macro-differentials of all of the molecules. The next step was the employment of these concepts to various applications.

In the first application, Planck considered a gas in a state of equilibrium and derived the law of distribution of the velocities of the molecules. The volume of the gas is

$$V = \int dx \cdot dy \cdot dz,$$

and the total energy

$$E = \frac{m}{2} \sum \left(\dot{x}^2 + \dot{y}^2 + \dot{z}^2 \right) f\sigma. \quad (18)$$

For a gas in equilibrium, $\delta S = 0$, and from equation (17),

$$\sum (\log f + 1) \cdot \delta f \cdot \sigma = 0, \quad (19)$$

where δf is a change in the law of distribution as a function of N, V and E.

The total number of molecules does not change, so in accordance with equation (14),

$$\sum \delta f \cdot \sigma = 0,$$

and

$$\sum \left(\dot{x}^2 + \dot{y}^2 + \dot{z}^2 \right) \cdot \delta f \cdot \sigma = 0$$

since the energy is constant and through the use of equation (18). The necessary and sufficient conditions for all δf to fulfill equation (19) is

$$\log f + \beta \left(\dot{x}^2 + \dot{y}^2 + \dot{z}^2 \right) = \text{const},$$

or

$$f = \alpha e^{-\beta (\dot{x}^2 + \dot{y}^2 + \dot{z}^2)},$$

which is the (well known) Maxwellian distribution of the velocities, where α and β are constants and the spatial distribution of the molecules is uniform.

The next step was to solve for the constants in the above equation by substitutions of *f* into equation (14) for the number of molecules in the volume, and into equation (18) for the energy. Then substituting into equation (17), the entropy of the gas is expressed in terms of N, V and E[6],

$$S = \text{const} + kN\left(\frac{3}{2}\log E + \log V\right), \quad (20)$$

where the constant contains terms in N and m.

Applying equation (20) and the general thermodynamic definition of entropy,

$$dS = \frac{dE + pdV}{T},$$

the partial differentials with respect to E and V are easily solved,

$$\left(\frac{dS}{dE}\right)_V = \frac{3}{2}\frac{kN}{E} = \frac{1}{T}, \quad (21),$$

$$\left(\frac{dS}{dV}\right)_E = \frac{kN}{V} = \frac{p}{T}, \quad (22)$$

resulting in

$$p = \frac{kNT}{V} = \frac{RnT}{V}$$

6 These substitutions are simple and straightforward, and only the result is shown. See the unabridged version of this chapters for the details.

which contains the laws of Boyle, Gay Lussac and Avogadro, and in which R is the absolute gas constant ($8.315 \cdot 10^7$ erg/deg). The constant, k, was then determined by substituting the values for the absolute gas constant, the number of mols of the gas ($O_2 = 32\ g$) and the mass of a molecule of gas,

$$k = \frac{Rn}{N}, \quad (23).$$

The ratio of the molecular mass to the mol mass is

$$\omega = \frac{n}{N},$$

resulting in

$$k = \omega R. \quad (24)$$

Re-arranging equation (21),

$$E = \frac{3}{2} kNT. \quad (25)$$

The energy of an ideal gas is

$$E = Anc_v T,$$

where c_v is the heat capacity, in calories, of a mol of constant volume, and A is the mechanical equivalent of heat,

$$A = 4.19 \cdot 10^7 \frac{\text{erg}}{\text{cal}},$$

and substituting from equations (23), (24) and (25),

$$c_v = \frac{3kN}{2An},$$

and

$$c_v = \frac{3}{2}\frac{R}{A} = 3.0, \quad (26).$$

which is the mol heat, in calories, of a monatomic gas at constant volume.

Applying the first law of thermodynamics,

$$c_p - c_v = \frac{R}{A},$$

where c_p is the mol heat at constant pressure. The mean kinetic energy L of a molecule is obtained by applying equation (26),

$$c_p = 5, \quad \frac{c_p}{c_v} = \frac{5}{3},$$

$$L = \frac{E}{N} = \frac{3}{2}kT. \quad (27)$$

The mechanical and thermodynamic expressions for the entropy were thus used to derive the above relations.

Planck pointed out that the above methods can be applied to other applications which are less easily handled, such as a monatomic gas that is not in an ideal state. In this case the finite size of atoms and the forces acting among them must be considered (J.D. van der Waals), which involves changes in the probability functions and the energy of the gas. The corresponding change in the condition for equilibrium agrees with that of van der Waals.

Planck took issue with the assumptions made by Boltzmann regarding polyatomic gases, arguing that the fact that the intra-molecular energy increases with temperature implies that the molecular heat of a polyatomic gas must be higher than that of a monatomic gas. The "law of uniform energy distribution is not applicable to intra-molecular processes". The laws of heat radiation, discussed in his next lecture, provide further contradictions, as he points out in a later chapter.

Planck closed this lecture by commenting that the principles that have been developed from the kinetic gas theory are not valid for the instable atoms of radioactive substances and this subject is left for future investigations.

The concept of entropy has been applied to many other applications in the succeeding years. The principle of *cybernetics* was based partly on entropy and probability theory (1948-1949), as was the initial development of *information theory* and new methods of analysis of noise in communications systems (1953-1964). The relationships between electrical noise and Planck's radiation theory, which lead to new and different pictures of radiation, are presented in detail in the final chapter of this book.

In spite of the fact that Planck's mathematics of probability was comparatively simple, the results of the analysis correlate with real world measurements and the laws of physics, just as was the case for the development of the thermodynamic equations of Lecture 2. These examples apply to both mechanical systems of gases. The next step is to accomplish a similar result for radiation.

Planck's Columbia Lectures

Chapter 5

Introduction to the Fifth Lecture

In the previous lecture, Planck applied his kinetic theory to the atoms in a gas and verified the results by conformance with measurements and the first and second laws of thermodynamics. Having covered the mechanical portion of his theory, he now proceeds to analyze the radiation of heat, as based on electromagnetic theory. The definitions of energy states and equilibrium are applied to the radiation process.

In contrast to Einstein's view of light as being in the form of "particles" (photons), heat rays are treated here as electromagnetic waves that are equivalent to light rays. Since energy is exchanged in finite amounts (quanta) by both theories, the primary difference between the two views is reduced to a question of the geometry of radiation.

Planck shows how radiation is related to entropy and probabilities and, in particular, temperature and the change in energy of the atoms. He takes great care in using the variables and parameters of the previous lectures wherever possible. This restriction adds to the depth of the analysis, and the results are very simple and clear.

Although some of the energy of a light ray striking the surface of a body is either radiated or reflected, Planck is mainly concerned with "black radiation", which applies to a restrictive condition in which the spectrum of radiation is quite broad (not coherent). He defines radiation in the form of an equations that he develops in a lengthy analytical process. His mathematical method also applies to an unique experiment by Wien, wherein black radiation from a heated enclosure was accurately measured.

The last part of this lecture covers electromagnetic theory and its relationship to elementary radiating and emitting bodies.

FIFTH LECTURE.

HEAT RADIATION. ELECTRODYNAMIC THEORY.

Last week I endeavored to point out that we find in the atomic theory a complete explanation for the whole content of the two laws of thermodynamics, if we, with Boltzmann, define the entropy by the probability, and I have further shown, in the example of an ideal monatomic gas, how the calculation of the probability, without any additional special hypothesis, enables us not only to find the properties of gases known from thermodynamics, but also to reach conclusions which lie essentially beyond those of pure thermodynamics. Thus, e. g., the law of Avogadro in pure thermodynamics is only a definition, while in the kinetic theory it is a necessary consequence; furthermore, the value of c_v, the mol-heat of a gas, is completely undetermined by pure thermodynamics, but from the kinetic theory it is of equal magnitude for all monatomic gases and, in fact, equal to 3, corresponding to our experimental knowledge. Today and tomorrow we shall be occupied with the application of the theory to radiant heat, and it will appear that we reach in this apparently quite isolated domain conclusions which a thorough test shows are compatible with experience. Naturally, we take as a basis the electro-magnetic theory of heat radiation, which regards the rays as electro-magnetic waves of the same kind as light rays.

We shall utilize the time today in developing in bold outline the important consequences which follow from the electro-magnetic theory for the characteristic quantities of heat radiation, and tomorrow seek to answer, through the calculation of the entropy, the question concerning the dependence of these quantities upon the temperature, as was done last week for ideal gases. Above all, we are concerned here with the determination of those quantities which at any place in a medium traversed by heat rays determine the state of the radiant heat. The state of radiation at a given place will not be represented by a vector which is determined by three components; for the energy flowing in a given direction is quite independent of that flowing in any other direction. In order to know the `state of radiation, we must be able to specify, moreover, the energy which in the time dt flows through a surface element $d\sigma$ for every direction in space. This will be proportional to the magnitude of $d\sigma$, to the time dt, and to the cosine of the angle ϑ which the direction considered makes with the normal to $d\sigma$. But the quantity to be multiplied by $d\sigma \cdot dt \cdot \cos\vartheta$ will not be a finite quantity; for since the radiation through any point of $d\sigma$ passes in all directions, therefore the quantity.will also depend upon the magnitude of the solid angle $d\Omega$, which we shall assume as the same for all points of $d\sigma$. In this manner we obtain for the energy which in the time dt flows through the surface element $d\sigma$, in the direction of the elementary cone $d\Omega$, the expression:

$$K\, d\sigma\, dt \cdot cos\, \vartheta \cdot d\Omega. \qquad (28)$$

K is a positive function of place, of time and of direction, and is for unpolarized light of the following form:

$$K = 2 \int_0^\infty \Re_\nu \, d\nu \qquad (29)$$

where ν denotes the frequency of a color of wave length λ and whose velocity. of propagation is q:

$$\nu = \frac{q}{\lambda},$$

and $\mathfrak{R}_\nu$ denotes the corresponding intensity of spectral radiation of the plane polarized light.

From the value of K is to be found the space density of radiation ϵ, i. e., the energy of radiation contained in unit volume. The point 0 in question forms the centre of a sphere whose radius r we take so small that in the distance r no appreciable absorption of radiation takes place. Then each element $d\sigma$ of the surface of the sphere furnishes, by virtue of the radiation traversing the same, the following contribution to the radiation density at 0:

$$\frac{d\sigma \cdot dt \cdot K \cdot d\Omega}{r^2\, d\Omega \cdot q dt} = \frac{d\sigma \cdot K}{r^2\, q}.$$

For the radiation cone of solid angle $d\Omega$ proceeding from a point of $d\sigma$ in the direction toward 0 has at the distance r from $d\sigma$ the cross-section $r^2 d\Omega$ and the energy passing in the time dt through this cross-section distributes itself along the distance qdt. By integration over all of the surface elements $d\sigma$ we obtain the total space density of radiation at 0:

$$\epsilon = \int \frac{d\sigma\, K}{r^2\, q} = \frac{1}{q}\int K\, d\Omega,$$

wherein $d\Omega$ denotes the solid angle of an elementary cone whose vertex is 0. For uniform radiation we obtain:

$$\epsilon = \frac{4\pi K}{q} = \frac{8\pi}{q} \cdot \int_0^\infty \mathfrak{R}_\nu \, d\nu \,. \qquad (30)$$

The production of radiant heat is a consequence of the act of emission, and its destruction is the result of absorption. Both processes, emission and absorption, have their origin only in material particles, atoms or electrons, not at the geometrical bounding surface; although one frequently says, for the sake of brevity, that a surface element emits or absorbs. In reality a surface element of a body is a place of entrance for the radiation falling upon the body from without and which is to be absorbed; or a place of exit for the radiation emitted from within the body and passing through the surface in the outward direction. The capacity for emission and the capacity for absorption of an element of a body depend only upon its own condition and not upon that of the surrounding elements. If, therefore, as we shall assume in what follows, the state of the body varies only with the temperature, then the capacity for emission and the capacity for absorption of the body will also vary only with the temperature. The dependence upon the temperature can of course be different for each wave length.

We shall now introduce that result following from the second law of thermodynamics which will serve us as a basis in all subsequent considerations: " a system of bodies at rest of arbitrary nature, form and position, which is surrounded by a fixed shell impervious to heat, passes in the course of time from an arbitrarily chosen initial state to a permanent state in which the temperature of all bodies of the system is the same." This is the thermodynamic state of equilibrium in which the entropy of the system, among all those values which it may assume compatible with the total energy specified by the initial conditions, has a maximum value. Let us now apply this law to a single homogeneous isotropic medium which is of great extent in all directions of space and which, as in all cases subsequently considered, is surrounded by a fixed shell, perfectly reflecting as regards heat rays. The medium possesses for each frequency ν of the heat rays a finite capacity for emission and a finite capacity for absorption. Let us consider, now, such regions of the medium as are very far removed from the surface. Here the influence of the surface will be in any case vanishingly small, because no rays from the surface reach these regions, and on account of the homogeneity and isotropy of the medium we must conclude that the heat radiation is in thermodynamic equilibrium everywhere and has the same properties in all directions, so that $\mathfrak{R}_\nu$ the specific intensity of radiation of a plane polarized ray, is independent of the frequency ν, of the azimuth of polarization, of the direction of the ray, and of location. Thus, there will correspond to each diverging bundle of rays in an elementary cone $d\Omega$, proceeding from a surface element $d\sigma$, an exactly equal bundle oppositely directed, within the same elemental cone converging toward the surface element. This law retains its validity, as a simple consideration shows, right up to the surface of the medium, for in thermodynamic equilibrium each ray must possess exactly the same intensity *as* that of the directly opposite ray, otherwise, more energy would flow in one direction than in the opposite direction. Let us fix our attention upon a ray proceeding inwards from the surface, this must have the same intensity as that of the directly opposite ray coming from within, and from this it follows immediately that the state of radiation of the medium at all points on the surface is the same as that within. The nature of the bounding surface and the special extent of the medium are immaterial, and in a stationary state of radiation $\mathfrak{R}_\nu$ is completely determined *by* the nature o€ the medium for each temperature.

This law suffers a modification, however, in the special case that the medium is absolutely diathermanous for a definite frequency ν. It is then clear that the capacity for absorption and also that for emission must be zero, because otherwise no stationary state of radiation could exist, i.e., a medium emits no color which it does not absorb. But equilibrium can then obviously exist for every intensity of radiation of the frequency considered, i.e., $\mathfrak{R}_\nu$ is now undetermined and cannot be found without knowledge of the initial conditions. An important example of this is furnished by an absolute vacuum, which is diathermanous for all frequencies. In a complete vacuum thermodynamic equilibrium can therefore exist for each arbitrary intensity of radiation and for each frequency, i.e., for each arbitrary distribution of the spectral energy. From a general thermodynamic point of view this indeterminateness of the properties of thermodynamic states of equilibrium is explained through the presence of numerous different relative maxima of the entropy, as in the case of a vapor which is in a state of supersaturation. But among all the different maxima there is a special maximum, the absolute, which indicates stable equilibrium. In fact, we shall see that in a diathermanous medium for each temperature there exists a quite definite intensity of radiation, which is designated as the stable intensity of radiation of the frequency ν considered. But for the present we shall assume for all frequencies a finite capacity for absorption and for emission.

We consider now two homogeneous isotropic media in thermo-dynamic equilibrium separated from each other by a plane surface. Since the equilibrium will not be disturbed if one imagines for the moment the surface of separation between the two substances to be replaced by a surface quite non-transparent to heat radiation, all of the foregoing laws hold for each of the

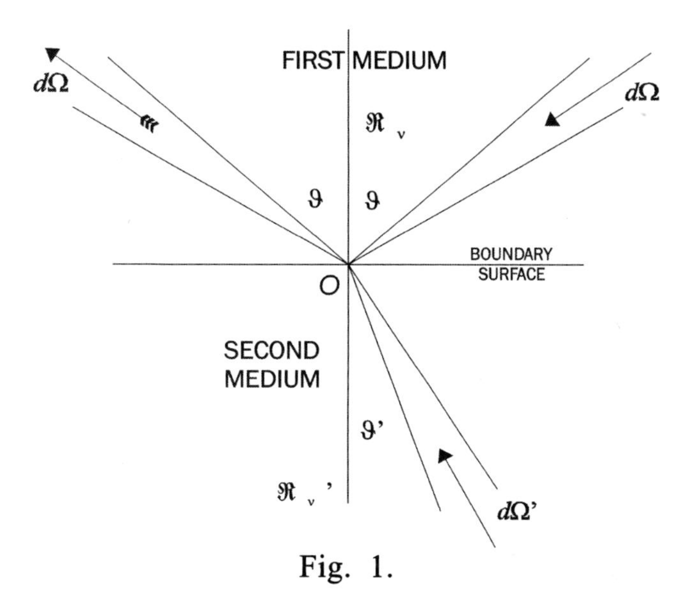

Fig. 1.

two substances individually. Let the specific intensity of radiation of frequency ν, polarized in any arbitrary plane within the first substance (the upper in Fig. 1)[1], be $\mathfrak{R}_\nu$ and that within the second substance $\mathfrak{R}_\nu'$ (we shall in general designate with a dash those quantities which refer to the second substance). Both quantities $\mathfrak{R}_\nu$ and $\mathfrak{R}_\nu'$, besides depending upon the temperature and the frequency, depend only upon the nature of the two substances, and, in fact, these values of the intensity of radiation hold quite up to the boundary surface between the substances, and are therefore independent of the

1 From my lectures upon the theory of heat radiation (Leipzig, J. A. Barth), wherein are to be found the details of the above somewhat abbreviated calculations.

properties of this surface.

Each ray from the first medium is split into two rays at the boundary surface: the reflected and the transmitted. The directions of these two rays vary according to the angle of incidence and the color of the incident ray, and, in addition, the intensity varies according to its polarization. If we denote by ρ (the reflection coefficient) the amount of the reflected energy of radiation and consequently by $1 - \rho$ the amount of transmitted energy with respect to the incident energy, then ρ depends upon the angle of incidence, upon the frequency and upon the polarization of the incident ray. Similar remarks hold for ρ', the reflection coefficient for a ray from the second medium, upon meeting the boundary surface.

Now the energy of a monochromatic plane polarized ray of frequency ν proceeding from an element $d\sigma$ of the boundary surface within the elementary cone $d\Omega$ in a direction toward the first medium (see the feathered arrow at the left in Fig. 1) is for the time dt, in accordance with (28) and (29):

$$dt \quad ds \quad \cos\vartheta \quad d\Omega \quad \mathfrak{R}\nu \quad d\nu, \qquad (31)$$

where

$$d\Omega \;=\; \sin\vartheta \; d\vartheta \; d\varphi. \qquad (32)$$

This energy is furnished by the two rays which, approaching the surface from the first and the second medium respectively, are reflected and transmitted respectively at the surface element ds in the same direction. (See the unfeathered arrows. The surface element $d\sigma$ is indicated only by the point 0.) The first ray proceeds in accordance with the law of reflection within the symmetrically drawn elementary cone $d\Omega$: the second approaches the surface within the elementary cone

$$d\Omega' = \sin\vartheta' d\vartheta' d\varphi', \qquad (33)$$

where, in accordance with the law of refraction,

$$\varphi' = \varphi \quad \text{and} \quad \frac{\sin\vartheta}{\sin\vartheta'} = \frac{q}{q'}. \qquad (34)$$

We now assume that the ray is either polarized in the plane of incidence or perpendicular to this plane, and likewise for the two radiations out of whose energies it is composed. The radiation coming from the first medium and reflected from $d\sigma$ contributes the energy:

$$\rho \cdot dt \cdot d\sigma \cdot \cos\vartheta \cdot d\Omega \cdot \mathfrak{R}_\nu \, d\nu, \qquad (35)$$

and the radiation coming from the second medium and transmitted through $d\sigma$ contributes the energy:

$$(1-\rho') \cdot dt \cdot d\sigma \cdot \cos\vartheta' \cdot d\Omega' \cdot \mathfrak{R}_\nu' d\nu. \qquad (36)$$

The quantities dt, $d\sigma$, ν, and $d\nu$ are here written without the accent, since they have the same values in both media.

Adding the expressions (35) and (36) and placing the sum equal to the expression (31), we obtain:

$$\rho \cos\vartheta \, d\Omega \mathfrak{R}_\nu + (1-\rho') \cdot dt \cdot d\sigma \cdot \cos\vartheta' \cdot d\Omega' \cdot \mathfrak{R}_\nu' d\nu.$$

Now, in accordance with (34):

$$\frac{\cos\vartheta\, d\vartheta}{q} = \frac{\cos\vartheta' d\vartheta'}{q'},$$

and further, taking note of (32) and (33):

$$d\Omega' \cos\vartheta' = d\Omega \cos\vartheta \cdot \frac{q'^2}{q^2},$$

and it follows that:

$$\rho\, \mathfrak{R}_\nu + (1 - \rho') \frac{q'^2}{q^2} \mathfrak{R}_\nu' = \mathfrak{R}_\nu$$

or

$$\frac{\mathfrak{R}_\nu}{\mathfrak{R}_\nu'} \cdot \frac{q^2}{q'^2} = \frac{(1-\rho')}{(1-\rho)}.$$

In the last equation the quantity on the left is independent of the angle of incidence ϑ and of the kind of polarization, consequently the quantity upon the right side must also be independent of these quantities. If one knows the value of these quantities for a single angle of incidence and for a given kind of polarization, then this value is valid for all angles of incidence and for all polarizations. Now, in the particular case that the rays are polarized at right angles to the plane of incidence and meet the bounding surface at the angle of polarization,

$$\rho = 0 \quad \text{and} \quad \rho' = 0.$$

Then the expression on the right will be equal to 1, and therefore it is in general equal to 1, and we have always:

$$\rho = \rho', \quad q^2 \mathfrak{R}_\nu = q'^2 \mathfrak{R}_\nu'. \qquad (37)$$

The first of these two relations, which asserts that the coefficient of reflection is the same for both sides of the boundary surface, constitutes the special expression of a general reciprocal law, first announced by Helmholz, whereby the loss of intensity which a ray of given color and polarization suffers on its path through any medium in consequence of reflection, refraction, absorption, and dispersion is exactly equal to the loss of intensity which a ray of corresponding intensity, color and polarization suffers in passing over the directly opposite path. It follows immediately from this that the radiation meeting a boundary surface between two media is transmitted or reflected equally well from both sides, for every color, direction and polarization.

The second relation, (37), brings into connection the radiation intensities originating in both substances. It asserts that in thermodynamic equilibrium the specific intensities of radiation of a definite frequency in both media vary inversely as the square of the velocities of propagation, or directly as the squares of the refractive indices. We may therefore write

$$q^2 \mathfrak{R}_\nu = F(\nu, T),$$

wherein F denotes a universal function depending only upon ν and T, the discovery of which is one of the chief problems of the theory.

Let us fix our attention again on the case of a diathermanous medium. We saw above that in a medium surrounded by a non-transparent shell which for a given color is diathermanous equilibrium can exist for any given intensity of radiation of this color. But it follows from the second law that, among all the intensities of radiation, a definite one, namely, that corresponding to the absolute maximum of the total entropy of the system, must exist, which characterizes the absolutely stable equilibrium of radiation. We now see that this indeterminateness is eliminated by the last equation, which asserts that in thermodynamic equilibrium the product $q^2 \mathfrak{R}_\nu$ is a universal function. For it results immediately therefrom that there is a definite value of $\mathfrak{R}_\nu$ for every diathermanous medium which is thus differentiated from all other values. The physical meaning of this value is derived directly from a consideration of the way in which this equation was derived: it is that intensity of radiation which exists in the diathermanous medium when it is in thermodynamic equilibrium while in contact with a given absorbing and emitting medium. The volume and the form of the second medium is immaterial; in particular, the volume may be taken arbitrarily small.

For a vacuum, the most diathermanous of all media, in which the velocity of propagation $q = c$ is the same for all rays, we can therefore express the following law: The quantity

$$\mathfrak{R}_\nu = \frac{1}{c^2} F(\nu, T) \qquad (38)$$

denotes that intensity of radiation which exists in any complete vacuum when it is in a stationary state as regards exchange of radiation with any absorbing and emitting substance, whose amount may be arbitrarily small. This quantity $\mathfrak{R}_\nu$ regarded as a function of ν gives the so-called normal energy spectrum.

Let us therefore consider a vacuum surrounded by given emitting and absorbing substances of uniform temperature. Then, in the course of time, there is established therein a normal energy radiation $\mathfrak{R}_\nu$ corresponding to this temperature. If now ρ_ν, be the reflection coefficient of a wall for the frequency ν, then of the radiation $\mathfrak{R}_\nu$ falling upon the wall, the part $\rho_\nu \mathfrak{R}_\nu$ will be reflected. On the other hand, if we designate by *Ev* the emission; coefficient of the wall for the same frequency ν, the total radiation proceeding from the wall will be:

$$\rho_\nu \mathfrak{R}_\nu + E_\nu = \mathfrak{R}_\nu,$$

since each bundle of rays possesses in a stationary state the intensity $\mathfrak{R}_\nu$. From this it follows that:

$$\mathfrak{R}_\nu = \frac{E_\nu}{1-\rho_\nu}, \qquad (39)$$

i.e., the ratio of the emission coefficient E_ν to the capacity for absorption $(1 - \rho_\nu)$ of a given substance is the same for all substances and equal to the normal intensity of radiation for each frequency (Kirchoff). For the special case that rv is equal to 0, i.e., that the wall shall be perfectly black, we have:

$$\mathfrak{R}_\nu = E_\nu,$$

that is, the normal intensity of radiation is exactly equal to the emission coefficient of a black body. Therefore the normal radiation is also called "black radiation". Again, for any given body, in accordance with (39), we have:

$$E_\nu < \mathfrak{K}_\nu,$$

i.e., the emission coefficient of a body in general is smaller than that of a black body. Black radiation, thanks to W. Wien and O. Lummer, has been made possible of measurement, through a small hole bored in the wall bounding the space considered.

We proceed now to the treatment of the problem of determining the specific intensity $\mathfrak{K}_\nu$ of black radiation in a vacuum, as regards its dependence upon the frequency ν and the temperature T. In the treatment of this problem it will be necessary to go further than we have previously done into those processes which condition the production and destruction of heat rays; that is, into the question regarding the act of emission and that of absorption. On account of the complicated nature of these processes and the difficulty of bringing some of the details into connection with experience, it is certainly quite out of the question to obtain in this manner any reliable results if the following law cannot be utilized as a dependable guide in this domain: a vacuum surrounded by reflecting walls in which arbitrary emitting and absorbing bodies are distributed in any given arrangement assumes in the course of time the stationary state of black radiation, which is completely determined by a single parameter, the temperature, and which, in particular, does not depend upon the number, the properties and the arrangement of the bodies. In the investigation of the properties of the state of black radiation the nature of the bodies which are supposed to be in the vacuum is therefore quite immaterial, and it is certainly immaterial whether such bodies actually exist anywhere in nature, so long as their existence and their properties are compatible throughout with the laws of electrodynamics and of thermodynamics. As soon as it is possible to associate with any given special kind and arrangement of emitting and absorbing bodies a state of radiation in the surrounding vacuum which is characterized by absolute stability, then this state can be no, other than that of black radiation. Making use of the freedom furnished by this law, we choose among all the emitting and absorbing systems conceivable, the most simple, namely, a single, oscillator at rest, consisting of two poles charged with equal quantities of electricity of opposite sign which are movable relative to each other in a fixed straight line, the axis of the, oscillator. The state of the oscillator is completely determined by its moment $f(t)$; i.e., by the product of the electric charge of the pole on the positive side of the axis into the distance between the poles, and by its differential quotient with regard to the time:

$$\frac{d\,f(t)}{d\,t} = \dot{f}(t),$$

The energy of the oscillator is of the following simple form:

$$U = \tfrac{1}{2} K f^2 + \tfrac{1}{2} L \dot{f}^2, \quad (40)$$

wherein K and L denote positive constants which depend upon the nature of the oscillator in some manner into which we need not go further at this time.

If, in the vibrations of the oscillator, the energy U remain absolutely constant, we should have: $dU = 0$ or:

$$K\,f(t) + L\,\ddot{f}(t) = 0\,,$$

and from this there results, as a general solution of the differential equation, a pure periodic vibration:

$$f = C\cos(2\pi\,\nu_0\,t - \vartheta)\,,$$

wherein C and ϑ denote the integration constants and ν_0 the number of vibrations per unit of time:

$$\nu_0 = \frac{1}{2\pi}\sqrt{\frac{K}{L}}\,, \qquad (41)$$

Such an oscillator vibrating periodically with constant energy would neither be influenced by the electromagnetic field surrounding it, nor would it exert any external actions due to radiation. It could therefore have no sort of influence on the heat radiation in the surrounding vacuum.

In accordance with the theory of Maxwell, the energy of vibration U of the oscillator by no means remains constant in general, but an oscillator by virtue of its vibrations sends out spherical waves in all directions into the surrounding field and, in accordance with the principle of conservation of energy, if no actions from without are exerted upon the oscillator, there must necessarily be a loss, in the energy of vibration and, therefore, a damping of the amplitude of vibration is involved.

In order to find the amount of this damping we calculate the quantity of energy which flows out through a spherical surface with the oscillator at the center, in accordance with the law of Poynting. However, we may not place the energy flowing outwards in accordance with this law through the spherical surface in an infinitely small interval of time dt equal to the energy radiated in the same time interval from the oscillator. For, in general, the electromagnetic energy does not always flow in the outward direction, but flows alternately outwards and inwards, and we should obtain in this manner for the quantity of the radiation outwards, values which are alternately positive and negative, and which also depend essentially upon the radius of the supposed sphere in such manner that they increase toward infinity with decreasing radius---which is opposed to the fundamental conception of radiated energy. This energy will, moreover, be only found independent of the radius of the sphere when we calculate the total amount of energy flowing outwards through the surface of the sphere, not for the time element dt, but for a sufficiently large time. If the vibrations are purely periodic, we may choose for the time a period; if this is not the case, which for the sake of generality we must here assume, it is not possible to specify a priori any more general criterion for the least possible necessary magnitude of the time than that which makes the energy radiated essentially independent of the radius of the supposed sphere.

In this way we succeed in finding for the energy emitted from the oscillator in the time from t to $t+\mathfrak{T}$ the following expression:

$$\frac{2}{3c^3}\int_t^{t+\mathfrak{T}} \ddot{f}^2(t)\,dt.$$

If now, the oscillator be in an electromagnetic field which

has the electric component $\mathfrak{E}_z$ at the oscillator in the direction of its axis, then the energy absorbed by the oscillator in the same time is:

$$\int_t^{t+\vartheta} \mathfrak{E}_z \dot{f} \cdot dt.$$

Hence, the principle of conservation of energy is expressed in the following form:

$$\int_t^{t+\vartheta} \left(\frac{dU}{dt} + \frac{2}{3c^3} \ddot{f}^2 - \mathfrak{E}_z \dot{f} \right) dt = 0.$$

This equation, together with the assumption that the constant

$$\frac{4\pi^2 \nu_0}{3c^3 L} = \sigma \qquad (42)$$

is a small number, leads to the following linear differential equation for the vibrations of the oscillator:

$$Kf + L\ddot{f} - \frac{2}{3c^3} \dddot{f} = \mathfrak{E}_z. \quad (43)$$

In accordance with what precedes, in so far as the oscillator is excited into vibrations by an external field $\mathfrak{E}_z$ one may designate it as a resonator which possesses the natural period ν_0 and the small logarithmic decrement σ. The same equation may be obtained from the electron theory, but I have considered it an advantage to derive it in a manner independent of any hypothesis concerning the nature of the resonator.

Now, let the resonator be in a vacuum filled with stationary black radiation of specific intensity $\mathfrak{R}_\nu$. How, then, does the mean energy U of the resonator in a state of stationary vibration depend upon the specific intensity of radiation $\mathfrak{R}_{\nu_0}$ with the natural period ν_0 of the corresponding color? It is this question which we have still to consider today. Its answer will be found by expressing on the one hand the energy of the resonator U and on the other hand the intensity of radiation $\mathfrak{R}_{\nu_0}$ by means of the component $\mathfrak{E}_z$ of the electric field exciting the resonator. Now however complicated this quantity may be, it is capable of development in any case for a very large time interval, from $t = 0$ to $t = \mathfrak{T}$ in the Fourier's series:

$$\mathfrak{E}_z = \sum_{n=1}^{n=\infty} C_n \cos\left(\frac{2\pi\, nt}{\mathfrak{T}} - \vartheta_n\right), \quad (44)$$

and for this same time interval $\mathfrak{T}$ the moment of the resonator in the form of a Fourier's series may be calculated as a function of t from the linear differential equation (43). The initial condition of the resonator may be neglected if we only consider such times t as are sufficiently far removed from the origin of time $t = 0$.

If it be now recalled that in a stationary state of vibration the mean energy U of the resonator is given, in accordance with (40), (41) and (42), by:

$$U = K\bar{f}^2 = \frac{16\pi^4\nu_0^3}{3\sigma c^3}\bar{f}^2,$$

it appears after substitution of the value of f obtained from the differential equation (43) that:

$$U = \frac{3c^3}{64\pi^2\nu_0^2}\Im\bar{C}_{n0}^2, \qquad (45)$$

wherein C_{no}^2 denotes the mean value of C_n for all the series of numbers n which lie in the neighborhood of the value $\nu_0\Im$, i. e., for which $\nu_0\Im$ is approximately $= 1$.

Now let us consider on the other hand the intensity of black radiation, and for this purpose proceed from the space density of the total radiation. In accordance with (30), this is:

$$e = \frac{8\pi}{c}\int_0^\infty \Re_\nu d\nu = \frac{1}{8\pi}\left(\bar{\mathfrak{E}}_x^2 + \bar{\mathfrak{E}}_y^2 + \bar{\mathfrak{E}}_z^2 + \bar{\mathfrak{H}}_x^2 + \bar{\mathfrak{H}}_y^2 + \bar{\mathfrak{H}}_z^2\right), (46)$$

and therefore, since the radiation is isotropic, in accordance with (44):

$$\frac{8\pi}{c}\int_0^\infty \Re_\nu d\nu = \frac{3}{4\pi}\bar{\mathfrak{E}}_z^2 = \frac{3}{8\pi}\sum_{n=1}^{n=\infty} C_n^2 .$$

If we write $\Delta n/\mathfrak{T}$ on the left instead of $d\nu$, where Δn is ,a large number, we get:

$$\frac{8\pi}{c}\sum_{n=1}^{n=\infty}\mathfrak{R}_{\nu}\frac{\Delta n}{T}=\frac{3}{8\pi}\sum_{n=1}^{n=\infty}C_n^2,$$

and obtain then by "spectral" division of this equation:

$$\frac{8\pi}{c}\mathfrak{R}_{\nu_0}\frac{\Delta n}{\mathfrak{T}}=\frac{3}{8\pi}\sum_{n_0-(\Delta n/2)}^{n_0+(\Delta n/2)}C_n^2,$$

and, if we introduce again the mean value

$$\frac{1}{\Delta n}\cdot\sum_{n_0-(\Delta n/2)}^{n_0+(\Delta n/2)}C_n^2=\bar{C}_{n_0}^2,$$

we then get:

$$\mathfrak{R}_{\nu_0}=\frac{3c\mathfrak{I}}{64\pi^2}\cdot\bar{C}_{n_0},$$

By comparison with (45) the relation sought is now found:

$$\mathfrak{R}_{\nu_0}=\frac{\nu_0^2}{c^2}U, \qquad (47)$$

which is striking on account of its simplicity: and, in particular, because it is quite independent of the damping constant σ of the resonator.

This relation, found in a purely electrodynamic manner, between the spectral intensity of black radiation and the energy of a vibrating resonator will furnish us in the next lecture, with the aid of thermodynamic considerations, the necessary means of attack in deriving the temperature of black radiation together with the distribution of energy in the normal spectrum.

Planck's Columbia Lectures

Summary of the Fifth Lecture

In this lecture, Planck determined the basic relationships between frequency, energy and spectral radiation and presented a model of radiation. Nineteen equations were numbered, most important most of which will be discussed in this summary. Further details about the mathematical manipulations can be found in this lecture.

In determining the state of radiation for radiant heat, a basic assumption is that energy flowing in a given direction is not affected by heat flowing in another direction. Therefore, the state variables are not represented by vectors, and vector addition does not apply. However, the energy flowing through a surface element $d\sigma$ and over a time increment dt for the three dimensions of space must be specified. The radiation density also varies with the cosine of the radiation angle θ (notice that I have chosen to use a different font variable for clarity) normal to the surface over the solid angle $d\Omega$ in the direction of radiation. Therefore, the energy flowing through this elementary cone of radiation is

$$K d\sigma\, dt \cdot \cos\theta \cdot d\Omega, \quad (28)$$

where K is a positive function of space, time and direction. The light generated varies over a wide frequency range, and K is determined from the total radiation over this frequency range,

$$K = 2\int_0^\infty \mathfrak{R}_\nu \, d\nu \quad (29)$$

where ν is the frequency of radiation (note that f, rather than ν, is the contemporary common choice for the variable to represent frequency in many disciplines) and $\mathfrak{R}_v$ is the spectral radiation intensity. It is also important to note that Planck most properly chose to write this equation in terms of frequency rather than wavelength,

$$\lambda = \frac{q}{v} = \frac{q}{f}$$

where q is the speed of light, which is equal to c in a vacuum.

The radiation to and from an element $d\sigma$ at a point on the surface to a given radius is over of a solid angle, Ω, and the area at that radius is that of a the surface of a sphere. Planck determined the space density of radiation at that point on the surface by forming an equation for the contribution of the radiation of that element to the radiation density and then integrating over the solid angle. Then, substituting equation (29), he obtained

$$\epsilon = \frac{4\pi K}{q} = \frac{8\pi}{q} \cdot \int_0^\infty \mathfrak{R}_v \, dv \quad (30)$$

under the assumption of uniform radiation and no heat loss in the medium.

Applying the concept of thermodynamic equilibrium, he asserted that a system of bodies, surrounded by a fixed shell that is impervious to heat, eventually reaches a state in which all of the bodies are at the same temperature. The entropy of such a body has reached its maximum. In a stationary state of radiation, $\mathfrak{R}_v$ is determined entirely by the nature of the medium at each temperature.

In a medium that is completely diathermanous (transmissive of heat) at a given frequency, there is neither absorption nor emission. Therefore, equilibrium can exist at any finite radiation intensity, since quantum absorption and emission at a given frequency are fundamental to the idea of stationary states. A complete vacuum is diathermanous for arbitrary intensities of radiation and the distribution of the spectral energy. Therefore, the medium does not have the capability of emission or absorption.

In the next step, he applied the known laws of radiation to a system of two mediums separated by a boundary, illustrated by the diagram below.

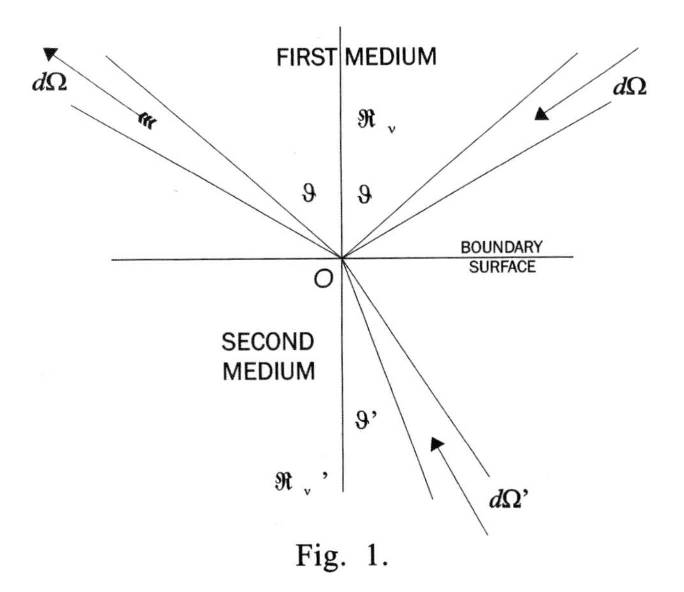

Fig. 1.

The angles of the reflected and refracted rays on both sides of the boundary are determined by the use of the law of refraction,

$$\varphi' = \varphi \quad \text{and} \quad \frac{\sin\vartheta}{\sin\vartheta'} = \frac{q}{q'}. \qquad (34)$$

The energy reflected from a surface element $d\sigma$ to the first medium is

$$\rho \cdot dt \cdot d\sigma \cdot \cos\vartheta \cdot d\Omega \cdot \mathfrak{R}_\nu \, d\nu \,, \qquad (35)$$

and the radiation from the second medium that is transmitted through $d\sigma$ is

$$(1-\rho') \cdot dt \cdot d\sigma \cdot \cos\vartheta' \cdot d\Omega' \cdot \mathfrak{R}_\nu ' d\nu . \qquad (36)$$

Using the above equations (32) through (36), the following relationship was determined:

$$\frac{\mathfrak{R}_\nu}{\mathfrak{R}_\nu '} \cdot \frac{q^2}{q'^2} = \frac{(1-\rho')}{(1-\rho)}.$$

The quantities on both sides of this equation are independent of the angle of incidence and the polarization. Choosing a particular case where the rays are polarized at right angles to the plane of incidence and intersect the boundary surface at the angle of polarization, the following simple relationship is obtained:

$$\rho = 0 \quad \text{and} \quad \rho' = 0.$$

The amplitude of the terms on both sides of the expression is then equal to one, which must then also apply to the general case, and therefore

$$\rho = \rho', \quad q^2 \mathfrak{R}_\nu = q'^2 \mathfrak{R}_\nu '. \quad (37)$$

The coefficient of reflection is the same for both sides of the boundary surfaces, and therefore radiation meeting a boundary surface between two media is transmitted or reflected equally well from both sides (first discovered by Helmholtz). In accordance with equation (37), the specific intensities of radiation at a given frequency vary in both media inversely with the square of the velocities of propagation or directly as the squares of the refractive indices (in thermodynamic equilibrium). Therefore,

$$q^2 \Re_\nu = F(\nu, T),$$

The problem now is to determine the term on the right of this equation in order to solve for Planck's radiation model equation. For absolutely stable equilibrium, the total entropy of the system must be a maximum. The term on the left of the above equation is a universal function, and therefore, the radiation function differ for each diathermanous medium in equilibrium.

For a vacuum, the velocity of propagation ($q = c$) is the same for all rays, and therefore

$$\Re_\nu = \frac{1}{c^2} F(\nu, T) \qquad (38)$$

denotes that the radiation intensity in a complete vacuum. This function is the normal energy spectrum with respect to frequency ν[7].

In the next step, a vacuum surrounded by emitting

7 This derivation begins with frequency as the variable in the expression, rather than the wavelength λ. A conversion process is necessary in order to transform the variables.

and absorbing bodies of uniform temperature was considered. With a reflection coefficient ρ_ν for the wall for a frequency ν and a radiation $\mathfrak{R}_\nu$ falling on the wall, the radiation from the wall is

$$\rho_\nu \mathfrak{R}_\nu + E_\nu = \mathfrak{R}_\nu,$$

where $\rho_\nu \mathfrak{R}_\nu$ is the reflected radiation and E_ν is the emission coefficient of the wall. Solving this equation results in

$$\mathfrak{R}_\nu = \frac{E_\nu}{1-\rho_\nu},$$

which is the ratio of the emission coefficient to the capacity of absorption. For black radiation the wall is perfectly black and the reflection coefficient is zero, resulting in

$$\mathfrak{R}_\nu = E_\nu,$$

so the normal intensity of radiation is equal to the emission coefficient of a black body. For all other bodies, the emission coefficient is less than that of a black body by equation (39). Black radiation was first measured (in an experiment conducted by W. Wien and O. Lummer) through a small hole bored in the wall of an enclosure.

In order to determine the specific intensity $\mathfrak{R}_\nu$ of black radiation in a vacuum, the following law is applied:

> For a vacuum that is completely surrounded by reflecting walls, black radiation is a function of only temperature and reaches a state of equilibrium in the course of time. The bodies from which the walls are constructed are immaterial except that their

properties must be compatible with the laws of electrodynamics and thermodynamics.

Planck chose the electronic oscillator, consisting of simply two poles charged with equal quantities of electricity that are movable with respect to each other along the axis of the oscillator, for the absorbing/reflecting body. The state of the oscillator is determined by its moment $f(t)$,which is the product of the electric charge of the pole on the positive side of he axis into the distance between the poles, and its differential with respect to time

$$\dot{f}(t) = \frac{df(t)}{dt}.$$

The energy of the oscillator is

$$U = \tfrac{1}{2}Kf^2 + \tfrac{1}{2}L\dot{f}^2, \qquad (40)$$

and the constants K and L depend upon the nature of the oscillator and will be determined later.

If the vibration of the oscillator is constant, then

$$dU = 0,$$

and therefore,

$$Kf(t) + L\ddot{f}(t) = 0.$$

The solution to this equation is a pure periodic vibration,

$$f(t) = C\cos(2\pi\nu_0 t - \vartheta),$$

wherein C and ϑ are integration constants and ν_0 is the oscillating frequency

$$\nu_0 = \frac{1}{2\pi}\sqrt{\frac{K}{L}}. \quad (41)$$

Such an oscillator, vibrating with constant frequency, would not be affected by and external oscillations due to radiation, and therefore it could have no influences on the heat radiation in the surrounding vacuum.

In accordance with Maxwell's theory, the energy U of the oscillator does not generally remain constant and sends out spherical waves in all direction into the surrounding field. If no actions from without are exerted upon the oscillator, then there must be a loss in energy, and the amplitude of the oscillations are damped. However, the energy generally flows both outward and inward in a manner that may, or may not, be periodic. The general criterion for the least possible necessary length of time is that which makes the energy radiated essentially independent of the radius of the sphere of radiation.

In this next step, the effect of an external field that excites the oscillator into vibrations by an external field $\mathfrak{E}_v$ is determined. This derivation can be obtained from electron theory, but Planck chose to derive it in a manner independent of any hypothesis concerning the nature of the resonator. The energy emitted from the oscillator in the time from t to $(t+T)$ is

$$\frac{2}{3c^3}\int_t^{t+\mathfrak{T}} \ddot{f}^2(t)dt.$$

If the oscillator is in an electric field with a component E_z

in the direction of its axis, then the energy absorbed by the oscillator in the same time is

$$\int_t^{t+\mathfrak{T}} \mathfrak{E}_z \dot{f} \cdot dt .$$

Applying the principle of conservation of energy under these conditions,

$$\int_0^{t+\mathfrak{T}} \left(\frac{dU}{dt} + \frac{2}{3c^3} \ddot{f}^2 - \mathfrak{E}_z \dot{f} \right) dt = 0 .$$

With the assumption that the constant

$$\sigma = \frac{4\pi^2 \nu_0}{3c^3 L} \quad (42)$$

is a small number, the following linear differential equation describes the vibrations of the oscillator:

$$Kf + L\ddot{f} - \frac{2}{3c^3} \dddot{f} = \mathfrak{E}_z . \quad (43)$$

This expression applies to an oscillator, at a frequency of ν_0, excited by a small logarithmic increment, into oscillations by an external field.

The resonator can be influenced by an external radiation field $\mathfrak{R}\nu_0$ (ν_0 is its natural period), in which case the mean energy U of the resonator, in a state of stationary vibration, varies accordingly. In determining this effect, the expressions for both U and $\mathfrak{R}\nu_0$ are to each be expressed as a function of the component E_z of the electric field exciting the resonator. For a very large time interval, from $t = 0$ to

$t = \mathfrak{J}$, the electric field may be expressed in the Fourier's series:

$$\mathfrak{E}_z = \sum_{n=1}^{n=\infty} C_n \cos\left(\frac{2\pi\, nt}{\mathfrak{J}} - \vartheta_n\right), \quad (44)$$

and, for this same time interval $\mathfrak{J}$, the moment of the resonator in the form of a Fourier's series may be calculated as a function of t from the linear differential equation (43). The initial condition of the resonator can be neglected for very large values of time t.

The mean energy U of the resonator is calculated using equations (40), (41) and (42):

$$U = K\,\overline{f}^2 = \frac{16\pi^4 \nu_0^3}{3\sigma c^3}\,\overline{f}^2,$$

and upon substituting the value of f obtained from the differential equation (43),

$$U = \frac{3c^3}{64\pi^2 \nu_0^2}\,\mathfrak{J}\,\overline{C}_{n0}^{\,2}, \quad (45)$$

wherein C_{no}^2 denotes the mean value of C_n in the neighborhood of $\nu_0\,\mathfrak{J}$, ($\nu_0\,\mathfrak{J}$ is approximately equal to one).

The space density of the total radiation, in accordance with black radiation and equation (30), is:

$$\epsilon = \frac{8\pi}{c}\int_0^\infty \mathfrak{R}_\nu d\nu = \frac{1}{8\pi}\left(\overline{\mathfrak{E}}_x^{\,2} + \overline{\mathfrak{E}}_y^{\,2} + \overline{\mathfrak{E}}_z^{\,2} + \overline{\mathfrak{H}}_x^{\,2} + \overline{\mathfrak{H}}_y^{\,2} + \overline{\mathfrak{H}}_z^{\,2}\right), (46)$$

and therefore, since the radiation is isotropic in accordance with (44),

$$\frac{8\pi}{c}\int_0^\infty \mathfrak{R}_\nu d\nu = \frac{3}{4\pi}\overline{\mathfrak{E}_z^{\,2}} = \frac{3}{8\pi}\sum_{n=1}^{n=\infty} C_n^{\,2}\,.$$

Substituting $\Delta n/\mathfrak{T}$ on the left instead of $d\nu$ (Δn is a large number),

$$\frac{8\pi}{c}\sum_{n=1}^{n=\infty}\mathfrak{R}_\nu\frac{\Delta n}{T} = \frac{3}{8\pi}\sum_{n=1}^{n=\infty} C_n^{\,2}\,,$$

and by "spectral" division of this equation:

$$\frac{8\pi}{c}\mathfrak{R}_{\nu_0}\frac{\Delta n}{\mathfrak{T}} = \frac{3}{8\pi}\sum_{n_0-(\Delta n/2)}^{n_0+(\Delta n/2)} C_n^{\,2}.$$

With the definition of the mean value as before,

$$\frac{1}{\Delta n}\cdot\sum_{n_0-(\Delta n/2)}^{n_0+(\Delta n/2)} C_n^{\,2} = \overline{C}_{n_0}^{\,2}\,,$$

then by substitution,

$$\mathfrak{R}_{\nu_0} = \frac{3c\mathfrak{J}}{64\pi^2}\cdot\overline{C}_{n_0}\,,$$

and using equation (45), the proper relation is determined:

$$\mathfrak{R}_{\nu_0} = \frac{\nu_0^2}{c^2} U, \quad (47)$$

which is independent of the damping constant σ of the resonator. This relation, between the spectral intensity of black radiation and the energy of a vibrating resonator, was determined in a pure electrodynamic manner. In the next lecture, thermodynamic considerations are used to derive the temperature of black radiation and the distribution of energy in the normal spectrum.

Chapter 6

Introduction to the Sixth Lecture

This is perhaps the most important and interesting of all the lectures. It is fascinating to see how Planck utilizes the foundation that he built in the earlier lectures and merges the various relationships to form his radiation model equation.

There is much information to absorb in the few pages that follow, and I found it valuable to proceed slowly in studying each step in his presentation of this lecture. One of the key elements upon which his theory is based, is the definition of the energy states of “resonators”. The relationships between probability theory and energy states is also important and leads to an understanding of the process of energy exchange between the resonators and the black radiation field.

The “secret of minus-one” is solved in this chapter, and readers will be able to judge as to whether or not his radiation equation was a simple empirical model. The mathematical terms that produce the “bending over of the curve” in the radiation equation, and the relationship of Stirling’s formula in deriving it, will also be found in this interesting chapter.

The commentary at the end of this chapter contains an analysis of the energy state equation and a graphical presentation that allows further insight as to the nature of electron motion within matter that is radiated by a black radiation field.

SIXTH LECTURE.

HEAT RADIATION. STATISTICAL THEORY.

Following the preparatory considerations of the last lecture we shall treat today the problem which we have come to recognize as one of the most important in the theory of heat radiation: the establishment of that universal function which governs the energy distribution in the normal spectrum. The means for the solution of this problem will be furnished us through the calculation of the entropy S of a resonator placed in a vacuum filled with black radiation and thereby excited into stationary vibrations. Its energy U is then connected with the corresponding, specific intensity $\mathfrak{K}_\nu$ and its natural frequency ν in the radiation of the surrounding field through equation (47):

$$\mathfrak{K}_\nu = \frac{\nu^2}{c^2} U . \qquad (48)$$

When S is found as a function of U, the temperature T of the resonator and that of the surrounding radiation will be given by:

$$\frac{dS}{dU} = \frac{1}{T}, \qquad (49)$$

and by elimination of U from the last two equations, we then find the relationship among $\mathfrak{K}_\nu$, T and ν.

In order to find the entropy S of the resonator we will utilize the general connection between entropy and probability, which we have extensively discussed in the previous lectures, and inquire then as to the existing probability that the vibrating resonator possesses the energy U. In accordance with what we have seen in connection with the elucidation of the second law through atomistic ideas, the second law is only applicable to a physical system when we consider the quantities which determine the state of the system as mean values of numerous disordered individual values, and the probability of a state is then equal to the number of the numerous, a priori equally probable, complexions which make possible the realization of the state. Accordingly, we have to consider the energy U of a resonator placed in a stationary field of black radiation as a constant mean value of many disordered independent individual values, and this procedure agrees with the fact that every measurement of the intensity of heat radiation is extended over an enormous number of vibration periods. The entropy of a resonator is then to be calculated from the existing probability that the energy of the radiator possesses a definite mean value U within a certain time interval.

In order to find this probability, we inquire next as to the existing probability that the resonator at, any fixed time possesses a given energy, or in other words, that that point (the state point) which through coordinates indicates the state of the resonator falls in a given "state domain. At the conclusion of the third lecture (p. 57 – of the actual manuscript) we saw in general that this probability is simply measured through the magnitude of the corresponding state domain:

$$\int d\varphi \cdot d\psi ,$$

in case one employs as coordinates of state the general coordinate φ and the corresponding momentum ψ.

Now in general, the energy of the resonator, in accordance with (40), is:

$$U = \tfrac{1}{2} K f^2 + \tfrac{1}{2} L \dot{f}^2 .$$

If we choose f as the general coordinate φ and put, therefore $\varphi = f$, then the corresponding impulse ψ is equal

$$\frac{\partial U}{\partial f} = L \dot{f} ,$$

and the energy U expressed as a function of φ and ψ is:

$$U = \frac{1}{2} K \varphi^2 + \frac{1}{2} \frac{\psi^2}{L} .$$

If now we desire to find the existing probability that the energy of a resonator shall lie between U and $U + \Delta U$, we have to calculate the magnitude of that state domain in the (φ, ψ)-plane which is bounded by the curves $U =$ const. and $U + \Delta U =$ const. These two curves are similar and similarly placed ellipses and the portion of surface bounded by them is equal to the difference of the areas of the two ellipses. The, areas are respectively U/ν , and $(U + \Delta U)/\nu$; consequently, the magnitude sought for the state domain is: $\Delta U/\nu$. Let us now consider the whole state plane so divided into elementary portions by a large number of ellipses, such that the annular areas between consecutive ellipses are equal to each other; i. e., so that:

$$\frac{\Delta U}{\nu} = \text{const} = h.$$

We thus obtain those portions ΔU of the energy which correspond to equal probabilities and which are therefore to be designated as the energy elements:

$$\epsilon = \Delta U = h\nu. \qquad (50)$$

If the determination of the elementary domains is effected in a manner quite similar to that employed in the kinetic gas theory, there exist; with respect to the relationships there found, very notable differences. In the first place, the state of the physical system considered here, the resonator, does not depend as there upon the coordinates and the velocities, but upon the energy only, and this circumstance' necessitates that the entropy of a state depend, not upon the distribution of the state quantities φ and ψ, but only upon the energy U. A further difference consists in this that we have to do in the case of molecules with special mean values as regards time. But this distinction may be disregarded when we reflect that the mean time value of the energy U of a given resonator is obviously identical with the mean space value at a given instant of time of a great number N of similar resonators distributed in the same stationary field of radiation. Of course these resonators must be placed sufficiently far apart in order not directly to influence one another. Then the total energy of all the resonators:

$$U_N = NU. \qquad (51)$$

is quite irregularly distributed among all the individual resonators, and we have referred back the disorder as regards time to a disorder as regards space.

We are now concerned with the probability W of the state determined by the energy U_N of the N resonators placed in the same stationary field of radiation; i.e., with the number of individual arrangements or complexions which correspond to the distribution of energy U_N among the N resonators. With this in view, we subdivide the given total energy U_N into its elements ϵ so that:

$$U_N = P\epsilon. \qquad (52)$$

These P energy elements are to be distributed in every possible manner among the N resonators. Let us consider, then, the N resonators to be numbered and the figures written beside one another in a series, and in such manner that the number of times each figure appears is equal to the number of energy elements which fall upon the corresponding resonator. Then we obtain through such a number series a representation of a fixed complexion, in which with each individual resonator there is associated a definite energy. For example, if there are $N = 4$ resonators and $P = 6$ energy elements present, then one of the possible complexions is represented by the number series

$$1 \ 1 \ 3 \ 3 \ 3 \ 4$$

which asserts that the first resonator contains two, the second 0, the third 3, and the fourth 1 energy element. The totality of numbers in the series is 6, equal to the number of the energy elements present. The arrangement of figures in the series is immaterial for any complexion, since the mere interchange of figures does not change. the energy of a given resonator. The number of all the possible different complexions is therefore equal to the number of possible “combinations with repetition” of 4 elements with 6 classes:

$$W = \frac{(4+6-1)!}{(4-1)!6!} = \frac{9!}{3!6!} = 84 \,,$$

or, in our general case the probability sought is:

$$W = \frac{(N+P-1)!}{(N-1)!P!} \,,$$

We obtain, therefore, for the entropy S_N of the resonator system, in accordance with equation (12), since N and P are large numbers,

$$S_N = k \log \frac{(N+P)!}{N!P!} \,,$$

and with the aid of Sterling's formula (16):

$$S_N = k! \{ (N+P) \log (N+P) - N \log N - P \log \mathrm{P} \} \,.$$

If, in accordance with (52), we now write U_N/ϵ for P, NU for U_N in accordance with (51), and $h\nu$ for ϵ, in accordance with (50), we obtain, after an easy transformation, for the mean entropy of a single resonator:

$$\frac{S_N}{N} = S = k\{(1+\frac{U}{h\nu})\log(1+\frac{U}{h\nu}) - \frac{U}{h\nu}\log\frac{U}{h\nu}\}$$

as the solution of the problem in hand.

We will now introduce the temperature T of the resonator, and will express through T the energy U of the resonator and also the intensity $\mathfrak{R}_\nu$ of the heat radiation related to it through a stationary state of energy exchange. For this purpose we utilize equation (49) and obtain then for the energy of the resonator:

$$U = \frac{h\nu}{e^{\frac{h\nu}{kT}} - 1}.$$

It is to be observed that we have not here to do with a uniform distribution of energy (cf. p. 68) among the various resonators.

For the specific intensity of the monochromatic plane polarized ray of frequency ν, we have, in accordance with (48):

$$\mathfrak{R}_\nu = \frac{h\nu^3}{c^2} \cdot \frac{1}{e^{\frac{h\nu}{kT}} - 1}. \qquad (53)$$

This expression furnishes for each temperature T the energy distribution in the normal spectrum of a black body. A comparison with equation (38) of the last lecture furnishes us then with the universal function:

$$F(\nu, T) = \frac{h\nu^3}{e^{\frac{h\nu}{kT}} - 1}.$$

If we refer the specific intensity of a monochromatic ray, not to the frequency ν, but, as is commonly done in experimental physics, to the wave length λ, then, since between the absolute values of $d\nu$ and $d\lambda$ the relation exists:

$$|d\nu| = \frac{c \cdot |d\lambda|}{\lambda^2},$$

we obtain from

$$E_\lambda |d\lambda| = \Re_\nu |d\nu|,$$

the relation:

$$E_\lambda = \frac{c^2 h}{\lambda^5} \cdot \frac{1}{e^{\frac{ch}{k\lambda T}} - 1} \qquad (54)$$

as the intensity of a monochromatic plane polarized ray of wave length λ which is emitted normally to the surface of a black body in a vacuum at temperature T. For small values of λT (54) reduces to:

$$E_\lambda = \frac{c^2 h}{\lambda^5} \cdot e^{-(\frac{ch}{k\lambda T})}, \qquad (55)$$

which expresses Wien's Displacement Law. For large values of λT on the. other hand, there results from (54):

$$E_\lambda = \frac{ckT}{\lambda^4}, \qquad (56)$$

a relation first established by Lord Rayleigh and which we may here designate as the Rayleigh Law of Radiation.

From equation (30), taking account of (53), we

obtain for the space density of black radiation in a vacuum:

$$M = \frac{48\pi h}{c^3}\left(\frac{kT}{h}\right)^4 \cdot \alpha = aT^4 ,$$

Wherein

$$\alpha = 1 + \frac{1}{2^4} + \frac{1}{3^4} + \frac{1}{4^4} + \cdot\cdot\cdot = 1.0823.$$

The Stefan-Boltzmann law is hereby expressed. In accordance with the measurements of Kurlbaum, we have the constant

$$a = \frac{48\pi k^4}{c^3 h^3} \cdot \alpha = 7.061 \cdot 10^{-15} \frac{\text{erg}}{\text{cm}^3 \deg^4} ,$$

For that wave length λ_m which corresponds in the spectrum of black radiation to the maximum intensity of radiation E_λ we have from equation (54):

$$\left(\frac{dE_\lambda}{d\lambda}\right)_{\lambda = \lambda_m} = 0.$$

Carrying out the differentiation, we get, after putting for brevity:

$$\frac{ch}{k\lambda_m T} = \beta , \quad e^{-\beta} + \frac{\beta}{5} - 1 = 0.$$

The root of this transcendental equation is:

$$\beta = 4.9651 ;$$

and $\lambda_m T = ch/k\beta = b$ is a constant (Wien's Displacement Law). In accordance with the measurements of O. Lummer and E. Pringsheim,

$$b = 0.294 \text{ cm - deg.}$$

From this there follow the numerical values

$$k = 1.346.10^{-16} \text{ erg/deg, and } h = 6.548.10^{-27} \text{ erg-sec}$$

The value found for k easily permits of the specification numerically, in the C.G.S. system, of the general connection between entropy and probability, as expressed through the universal equation (12). Thus, quite in general, the entropy of a physical system is

$$S = 1.346.\ 10^{-16} \log_e W.$$

In the application to the kinetic gas theory we obtain from equation (24) for the ratio of the molecular mass to the mol mass:

$$\infty = \frac{k}{R} = 1.62 \cdot 10^{-24},$$

i.e., to one mol there corresponds $1 / \infty = 6.175 \cdot 10^{23}$ molecules, where it is supposed that the mol of oxygen

$$O_2 = 32 \text{ g.}$$

Accordingly, the number of molecules contained in 1 cu. cm. of an ideal gas at 0° Cels. and at atmospheric pressure is:

$$N = 2.76 \cdot 10^{19} .$$

The mean kinetic energy of the progressive motion of a molecule at the absolute temperature $T = 1$ in the absolute C.G.S. system, in accordance with (27), is:

$$L = 3k/2 = 2.02 \cdot 10^{-16}.$$

In general, the mean kinetic energy of progressive motion of a molecule is expressed by the product of this number and the absolute temperature T.

The elementary quantum of electricity, or the free electric charge of a monovalent ion or electron, in electrostatic measure is:

$$e = \infty \cdot 9658 \cdot 3 \cdot 10^{10} = 4.69 \cdot 10^{-10} .$$

This result stands in noteworthy agreement with the results of the latest direct measurements of the electric elementary quantum made by E. Rutherford and H. Geiger, and E. Regener.

Even if the radiation formula (54) here derived had shown itself as valid with respect to all previous tests, the theory would still require an extension as regards a certain point; for in it the physical meaning of the universal constant h remains quite unexplained. All previous attempts to derive a radiation formula upon the basis of the known laws of electron theory, among which the theory of J. H. Jeans is to be considered as the most general and exact, have led to the conclusion that h is infinitely small so that, therefore, the radiation formula of

Rayleigh possesses general validity, but, in my opinion, there can be no doubt that this formula loses its validity for short waves, and that the pains which Jeans has taken to place[8] the blame for the contradiction between theory and experiment upon the latter are unwarranted.

Consequently, there remains only the one conclusion, that previous electron theories suffer from an essential incompleteness which demands a modification, but how deeply this modification should go into the structure of the theory is a question upon which views are still widely divergent. J. J. Thompson inclines to the most radical view, as do J. Larmor, A. Einstein, and with him I. Stark, who even believe that the propagation of electromagnetic waves in a pure vacuum does not occur precisely in accordance with the Maxwellian field equations, but in definite energy quanta $h\nu$. I am of the opinion, on the other hand, that at present it is not necessary to proceed in so revolutionary a manner, and that one may come successfully through by seeking the significance of the energy quantum by solely in the mutual actions with which the resonators influence one another.[9] A definite decision with regard to these important questions can only be brought about as a result of further experience.

8 In that the walls used in the measurements of hollow space radiations must be diathermanous for the shortest waves.

9 It is my intention to give a complete presentation of these relatations in Volume 31 of th Annalen der Physik.

Summary of the Sixth Lecture

Planck was able to derive his radiation equation in only five short pages at the beginning of this chapter, which is an extraordinary feat. The derivation is based on eight simple equations (12, 13,16, 21,40, 47), plus the equations for a state domain and for the probability, *W*, of prior chapters. The insightful nature of Planck's methods become clear as we step through this process.

His first step was to define the energy of "a resonator" in a black radiation field,

$$\int d\varphi \cdot d\psi ,$$

where φ , which is the state coordinate, and ψ is the state momentum, which is a function of the derivative of φ. Planck's approach was based on energy concepts, and he began by utilizing the energy equation for a resonator (40),

$$U = \frac{1}{2}Kf^2 + \frac{1}{2}L\dot{f}^2 .$$

Planck did not define the constants *K* and *L*, although he chose the electric dipole with translational motion for his example in the fifth lecture. The resonant frequency for his example is $\omega = \sqrt{K/L}$, in accordance with equation (41). For a translational resonant mechanical system, *K* is the spring constant and *L* is the mass *m*, while for an electrical circuit 1/*K* is the capacitance *C* and *L* is the inductance. He avoided potential discourse as to the type of resonator by holding strictly to energy concepts in deriving the differential equations for both the stable and irradiated oscillators. However, he commented that these equations can also be obtained from electron theory. The discussions of the final chapter will contain further commentary on the pros and cons of this approach.

The next step was to derive the momentum term and substitute it into equation (40). The partial derivative of $U(f, \dot{f})$ with respect to $\dot{f}$ is $L\dot{f} = \psi$, which he calls the "impulse function". Substituting the result into equation (40), the resulting energy state equation is

$$U = \frac{1}{2} K\varphi^2 + \frac{1}{2}\frac{\psi^2}{L}.$$

This is an arithmetic equation that represents an ellipse, and the area of the ellipse divided by the frequency is equal to the energy of the resonator. Since an ellipse is a closed curve, the energy of a given ellipse is constant for each energy state. The stable resonator is therefore an oscillator, which is the term that he used in referring to it. The ellipse can be pictured as a circle by the proper transformation of variables if so desired.

Planck then applied his earlier assertion that the resonators have randomly distributed energies of fixed quantum levels. The difference in energy between any two states is

$$\Delta U = hv,$$

which he called the "energy element". Each element has equal probability, and the total energy of all of the resonators is the sum of the average energy of the resonators,

$$U_N = NU.$$

According to Planck, the ellipses are "similar" and "similarly placed", and the differential area between consecutive ellipses is constant.

There are a number of steps that he omitted in the above analysis. Important information can be obtained by analyzing the general equation for an ellipse,

$$\frac{\varphi^2}{a^2} + \frac{\psi^2}{b^2} = 1,$$

using various techniques, which will now be done.
The plot of this equation is shown in Figure 6.1 for a given stable state:

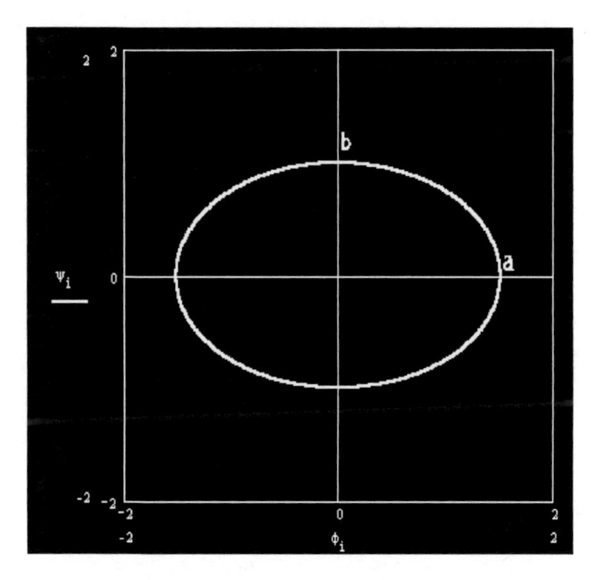

Figure 6.1 Plot of the State Space Equation

The values of a and b are determined by converting

the state equation into the ellipse equation, as shown above, and are found to be

$$a = \sqrt{\frac{2U}{K}} \quad b = \sqrt{2LU}.$$

The area of an ellipse is $A = \pi ab$, and upon substitution,

$$A = \pi\, ab = \pi \sqrt{\frac{4LU^2}{K}} = 2\pi U \sqrt{\frac{L}{K}}.$$

Then substituting the resonant frequency of equation (41),

$$A = 2\pi U \sqrt{\frac{L}{K}} = \frac{2\pi U}{\omega} = \frac{U}{v} = \pi\, ab.$$

Thus the area of the ellipse is equal to the energy of the resonator divided by the radian frequency, and the energy states are related to the parameters a and b of the equations of the ellipses.

For *stable* energy states, the state equation must form closed curves, and the ellipse is, of course, a closed curve. Planck asserted that there is a definite energy step that must take place between energy states and that the energy difference between adjacent states must be equal, since they are equally probable. Therefore, each energy element (state) is

$$\epsilon = (U + \Delta U) - \Delta U = \Delta U = \text{constant}.$$

This restriction limits the number of possible states, since the total energy is finite and there are N finite states,

$$U_N = NU.$$

The shapes of the ellipses vary with the relative variation in the values of the parameters a and b of the elliptical equation. Planck avoided considering the effects of these parameters, utilizing only the area of the curves and the energy state for his analysis. However, additional information may be obtained from a graphical analysis of energy state space, and several different possible scenarios for analysis will be illustrated. In the first case, it will be assumed that the two variables of the state equation vary in direct proportion, as shown in Figure 6.2.

U1

Figure 6.2 Quantum Energy States for Many Energy Levels (The circular ghost images are caused by the limited number of points in the plotting program and the graphics conversion method).

As the energy increases, the area of the circles increases proportionately. However, the variation in the parameters a

and b, which are proportional to the square root of the state energy U, become smaller in proportion at higher energy levels.

It is more likely that the state variables, φ and ψ , do not vary over equal ranges, as in Figure 6.3.

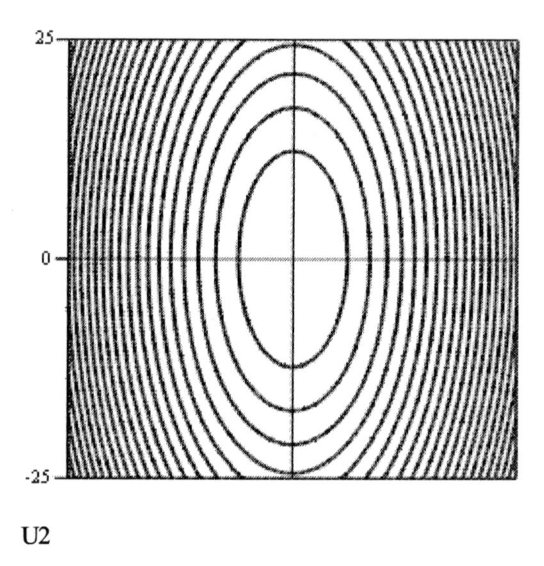

Figure 6.3 Plot of the State Equation for Unequal State Variables

In this instance, the state variables, φ and ψ, vary in proportion to each other, but their maximum values are not equal. Whether or not the actual energy states vary in this manner remains to be determined from other implications that follow from other physical restraints. For instance, if the position variable, φ, is limited in its allowable variation, then most of the energy change will take place in the direction of the momentum variable, ψ. The possible variations in the graphical shapes are unlimited, depending on the relationships between the two state variables. However, the additional possibilities will not be considered here. The plots of the quantum energy states, over a much greater range of energy levels, is shown below.

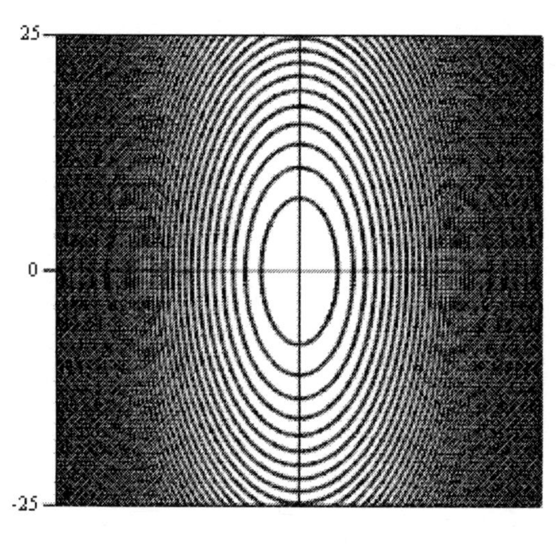

Figure 6.4 Plot of Many Energy States With Unequal State Variables.

For this plot, *b*>*a* and *b* is proportional to *a*, and therefore $\sqrt{KL} > 1$, where *K* and *L* are mechanical parameters. For an electrical system, the equivalent inequality is $\sqrt{L/C} > 1$, which is the impedance of the circuit. Therefore, the impedance of the system varies with the eccentricity of the ellipses in the above graph. For the system illustrated above, the impedance is constant for all energy states, since the state variables vary in proportion.

The definition of energy states, as based on the fixed difference in energy between successive states, can be analyzed in different ways. Planck did not utilize graphical analysis, but this method illustrates the parametric variations of many energy states under different scenarios. If the eccentricity of the ellipses does vary as a function of the system energy, then the impedance of the system not constant, which is another potential physical effect. Therefore, more information is obtained at the expense of additional effort. Other graphical methods will be employed later, as we attempt to better comprehend the implications of Planck's theory.

Planck next employed the concept of entropy, defined in the third lecture, which is a function of probable energy states. Each change in state involves a like energy change, and this energy corresponds to the mean time value for the energy of molecules. The same quantum principle applies to the radiation energy field of the molecules, and this field has a mean space value. In other words, these energies are mean values of the total energy that is distributed among a very large number of resonators. The total energy of the system,

$$U_N = NU,$$

is equal to the sum of the average energy of all of the resonators within a stationary field of radiation. The total energy is also the sum of the irregular distribution of the individual energies of the resonators, each of which is equally probable. In order to determine the probability distribution of the total energy, U_N, among the resonators, the first step is to determine the total number of allowable energy states, P. For a fixed energy element, ϵ,

$$U_N = P\epsilon.$$

The probability distribution is a function of the number of elements, N, and the allowable states, P, distributed in all possible "combinations with repetitions". Therefore,

$$W = \frac{(N+P-1)!}{(N-1)!P!},$$

where

$$N! \triangleq 1 \cdot 2 \cdot 3 \cdot 4 \cdot ....(N-1) \cdot N.$$

Note that the number of combinations of the N elements, without the order of the elements taken into consideration, is

$$C_N^P = \binom{N}{P} = \frac{N!}{P!(N-P)!},$$

and the equation for W is obtained by substituting $(N+P-1)$ for N in the above equation, which allows for the repetitions. Planck called W the "probability", which is a large number. Our present day definition of probability means *likelihood*, and is never greater than one and is the inverse of W.

At this point, Planck employed the entropy equation,

$$S_N = k \log W = k \log \frac{(N+P-1)!}{(N-1)!P!} \cong k \log \frac{(N+P)!}{N!P!}$$

wherein the approximation on the right applies for a large number of elements and states. Then, utilizing Stirling's equation as he did in the fourth lecture,

$$\log N! = N(\log N - 1),$$

and substituting,

$$S_N = k\left[(N+P)\log(N+P) - N\log N - P\log P\right].$$

This is the equation for the entropy of the system. The entropy of a single resonator is then,

$$S = \frac{S_N}{N} = k\left[(1+\frac{P}{N})\log(1+\frac{P}{N}) - \frac{P}{N}\log\frac{P}{N}\right].$$

In this equation, the entropy is expressed as a function of the number of resonators and the number of energy states. The total energy is the sum of all of the energies of the resonators. The minimum energy change is constant, and the average energy change is obtained from equation (52), $U_N = P\epsilon = NU$, so

$$\frac{P}{N} = \frac{U}{\epsilon} = \frac{U}{hv}.$$

Upon substitution into the entropy equation,

$$S = \frac{S_N}{N} = S = k\left[(1+\frac{U}{hv})\log(1+\frac{U}{hv}) - \frac{U}{hv}\log\frac{U}{hv}\right],$$

in which the entropy S is a function of the mean energy of the resonator.

The final step is to solve the entropy equation for the state energy, U, by applying the relationship between entropy and energy described by equation (49),

$$\frac{dS}{dU} = \frac{1}{T}.$$

The entropy equation can be differentiated directly and solved for *U*. In order to supply the steps that Planck omitted, a little arithmetic is required,

$$\Rightarrow \frac{1}{kT} = \left[\frac{1}{h\nu}\log(1+\frac{U}{h\nu}) - \frac{1}{h\nu}\log\frac{U}{h\nu}\right]$$

$$\Rightarrow \frac{h\nu}{kT} = \left[\log(1+\frac{U}{h\nu}) - \frac{1}{h\nu}\log\frac{U}{h\nu}\right] = \log\left[\frac{(1+\frac{U}{h\nu})}{\frac{U}{h\nu}}\right]$$

$$\Rightarrow e^{\frac{h\nu}{kT}} = \frac{(1+\frac{U}{h\nu})}{\frac{U}{h\nu}}$$

$$\Rightarrow U = \frac{h\nu}{\left(e^{\frac{h\nu}{kT}} - 1\right)}.$$

This state equation represents the average energy of a resonator in terms of frequency and temperature.

It is surprising that Planck did not number the above state energy equation, since it is very important. Also of importance, the spectral distribution of radiation is considerably different from the spectral distribution of the resonator energy. As Planck described it at the beginning of this lecture, the black radiation energy, $\mathfrak{R}_\nu$, of the surrounding field is "connected" with the resonator energy, *U*, through equation (47),

$$\mathfrak{R}_\nu = \frac{\nu^2}{c^2}U.$$

This relationship was determined in a purely electrodynamic manner in the fifth lecture. There are several basic considerations that were involved in deriving this equation.

The interface boundary between the radiator and the absorbing medium affects the radiation on both sides. The surface of the radiating medium can have a different coefficient of reflection, ρ, than that of the absorbing medium. A higher coefficient of reflection at the radiating surface will produce a lower level of radiation, which will therefore also affect the radiation equation. The transmitted energy is a function of the angle of incidence at the surface. Planck invoked the sine law of refraction, which is a function of the ratio of the ray angles on both sides of the interface, and this ratio is equal to the ratio of the velocities of propagation of the two mediums. This function applies to all angles of incidence and polarization, so Planck chose the simplest case, where the plane of incidence is at right angles to the plane of polarization. There is no reflected energy under these conditions, and the result is that the ratio of radiation levels is equal to the inverse of the ratio of velocities of propagation, q and q' of the two mediums,

$$q^2\mathcal{R}_\nu = q'^2\mathcal{R}'^2 = F(\nu,T),$$

where $F(\nu,T)$ is a general function that applies to either medium. In a vacuum, $q = c$, and

$$\mathfrak{R}_\nu = \frac{1}{c^2}F(\nu,T),$$

which is equation (38). The above steps have been repeated in order to clarify the methods that Planck used to establish the foundation of his theory.

Substituting the energy state equation into the above equation results in the spectral distribution of radiated energy,

$$\mathfrak{R}_{\nu} = \frac{h\nu^3}{c^2} \cdot \frac{1}{(e^{\frac{h\nu}{kT}} - 1)}.$$

The radiation process alters the spectral distribution of the energy state equation, as illustrated in the graphs of the two figures shown below.

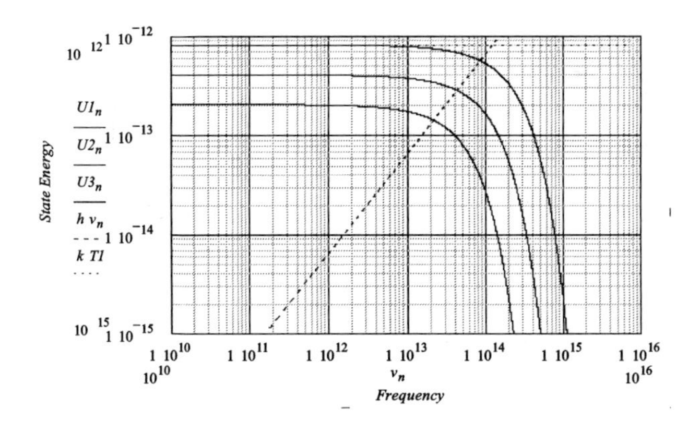

Figure 6.5. State Energy Spectral Distribution at Three Temperatures (6000, 3000,1500 deg K)

Figure 6.6. Radiated Energy Spectral Distribution at Three Temperatures (6000,3000, 1500 deg K)

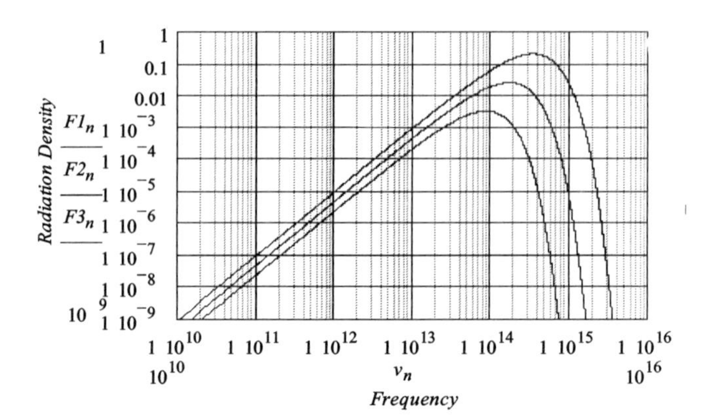

In Figure 6.5, the flat dotted line is a plot of kT, and the dashed line is a plot of hv. These plots intersect at similar points on the three temperature-related curves. There is an interesting significance associated with the points of intersection, at which $hv = kT$, that will be discussed in chapter nine.

The difference between the spectral distributions of these two figures clearly illustrates the differences between observed radiation and actual energy states. The two sets of curves cross over at the frequency, $v = c$, due to the effect of the v^2/c^2 term in the radiation equation. The importance of this term with respect to the radiation process will be discussed in Chapter 10. In comparing the units of the ordinates of the two graphs, the radiated energy appears to exceed the state energy, which is not the case. This difference is produced by the effect of the different units of measurements for two different process. Radiation occurs in rays, and these rays spread out with distance, whereas the state energy is a function of the movements of the molecules over a very small distance.

The final step in the derivation of the final form of Planck's radiation equation is comparatively straightforward. Optical measurements are most easily obtained by the measurement of wavelength, λ, rather than frequency. Wavelength is an inverse function of frequency, and so the wavelength equation is differentiated in order to obtain an expression in terms of differentials,

$$|d\nu| = \frac{c \cdot |d\lambda|}{\lambda^2}.$$

Substituting the wavelength differential into the frequency radiation equation,

$$E_\lambda = \frac{c^2}{\lambda^5} \frac{1}{\left(e^{\frac{ch}{k\lambda T}} - 1\right)},$$

which is the radiation equation (54) in this lecture.

The "minus-1", in the denominator of equations (53) and (54) and the above equation, was thus obtained, which is obviously not a mere substitution. It was derived beginning with the probability function, *W*, for the number of states, which was substituted into the entropy equation. An approximation for a large number of states resulted in a simplification of the expression. The derivative of this equation yields the energy state equation. The application of Stirling's equation was the key to finding the solution, and the –1 term appears in the denominator of the state energy equation. It seems like a lot of work to go through to find a simple constant, but it worked out very well.

Electromagnetic theory was used to determine the relationship between the energy of the resonator and the specific intensity of radiation. The effects produced at the interface boundary between the radiating medium and free space was determined, and from this information, the radiation equation, as a function of frequency, v, was obtained.

Planck converted the radiation equation into a function of wavelength rather than frequency, apparently because optical measurements are most easily made by measuring wavelength. In the analyses that have been added in this book, the frequency equations are utilized, since Fourier analysis (a very useful tool) is based on frequency transformations.

Planck then showed that at long wavelengths, his equation reduces to Rayleigh's Law of Radiation,

$$E_\lambda = \frac{ckT}{\lambda^4},$$

while at short wavelengths it reduces to Wien's Displacement Law,

$$E_\lambda = \frac{c^2 h}{\lambda^5} \cdot e^{-\left(\frac{ch}{k\lambda T}\right)},$$

thus confirming the accuracy of his radiation model at both ends of the spectrum.

In the fifth lecture, the space density of a cone of radiation was expressed in the form of an integral equation. This equation was solved by inserting the radiation equation into the integral, obtaining

$$\epsilon = \frac{8\pi}{q} \cdot \int_0^\infty \mathfrak{R}_v dv = \frac{48\pi h}{c^3} \left(\frac{kT}{h}\right)^4 \cdot \alpha = aT^4,$$

which is the Stefan-Boltzmann Law. Therefore, the quantum energy changes vary with the fourth power of temperature. Relating the measurements of Kurlbaum to this equation, he was able to solve for the constant,

$$a = 7.061 \cdot 10^{-15} \frac{\text{erg}}{\text{cm}^3\,\text{deg}^4}.$$

This equation shows that a quantum change in state energy produces a change in the in the intensity in a spatial cone of black radiation.

The final step in this process was the derivation of the constants, *k* and *h*. A third point on the energy curve is needed to make this determination. At the wavelength at which the radiation intensity is a maximum, the change in intensity is zero,

$$\left(\frac{dE_\lambda}{d\lambda}\right) = 0.$$

Upon differentiation,

$$e^\beta + \frac{\beta}{5} - 1 = 0,$$

which is a transcendental equation. The solution of this equation is

$$\beta = 4.9651 = \frac{ch}{k\lambda_m T}.$$

Therefore,

$$\lambda_m T = \frac{ch}{k\beta} = b = \text{const},$$

which is in accordance with Wien's Displacement Law. Again applying the results of real measurements (by O. Lummer and E Prinsheim),

$b = 0.294$ cm-deg,

resulting in

$h = 6.548 \cdot 10^{-27}$ erg-sec (6.62517 is the correct value),

and

$k = 1.346 \cdot 10^{-16}$ erg/deg (1.38044 is the correct value).

Planck's lengthy methodological approach has finally yielded the value of his now famous "Planck's constant", *h*, along with Boltzmann's constant.

The entropy of a physical system is

$$S = 1.346 \cdot 10^{-16} \log W,$$

upon substitution of Boltzmann's constant. Applying Boltzmann's constant to kinetic gas theory,

$$\infty = \frac{k}{R} = 1.62 \cdot 10^{-24},$$

and

$$N = \frac{1}{\infty} = 6.175 \cdot 10^{23} \text{ molecules/gm-mol}$$

(6.02486 is the correct value)

is the number of molecules in a gm-mol of molecules. In the example that he provided, the number of molecules of oxygen in a volume of one cubic cm under ideal conditions was calculated.

He provided other examples of the application of Boltzmann's constant to the determination of physical parameters. The mean kinetic energy of a molecule at one degree Kelvin by substituting the value of Boltzmann's constant into equation (27), from which the mean kinetic energy at any temperature can be determined by multiplying by the temperature. The value of the charge of an electron, in electrostatic units was obtained in a similar manner.

Planck questioned the physical meaning of the constant, h, and said that his theory would be extended by such a determination. In chapter nine, some insight as to the meaning of h will be presented.

The last paragraph is quite interesting, in that Planck brought up the question of whether or not Maxwell's electromagnetic field equations apply *precisely* to the "propagation of electromagnetic waves in a pure vacuum". Other famous scientists believed that radiation occurs in energy quanta, $h\nu$, and Einstein went so far as to define these quanta as "particles" that he called "photons". The evidence accumulated over the better part of a century, since the time of Planck, shows that Maxwell's equations are quite accurate in the analysis and measurement of low-frequency radiation. On the other hand, there is other evidence that tends to support Einstein's photon theory at optical frequencies and above. This apparent contradiction will be discussed in Chapter 10.

Planck's Columbia Lectures

Chapter 7

Introduction to the Seventh Lecture

Planck has covered the development of his famous radiation equation in great detail in the previous lectures, and we now know how he derived his radiation model and the "secret of minus-one". The main subject of this lecture is the application of his state space methods to the Principle of Least Action, as defined by Helmholtz. He believed that this principle would be fundamental to the unification of the "system of theoretical physics", since it connects together the elementary quantities by which physical phenomena are measured and from which physical laws can be interpreted. The integral equation (57) of this lecture, which expresses the Principle of Least Action, is fundamental to the analyses that follow.

There are a comparatively large number of equations that Planck uses in this development. However, this theory is important and extends over a wide range of applications in the analysis of dynamic systems, both mechanical and electrical.

Planck chooses two different sets of applications to which he applies the Principle of Least Action. In the first case, the system state coordinates are finite in number, while in the second case the coordinates form a continuous manifold. A surprising conclusion is reached at the end of this chapter - - - one that is so important that it could lead to the fundamental knowledge of the structure of the universe.

SEVENTH LECTURE

GENERAL DYNAMICS PRINCIPLE OF LEAST ACTION.

Since I began three weeks ago today to depict for you the present status of the system of theoretical physics and its probable future development, I have continually sought to bring out that in the theoretical physics of the future the most important and the final division of all physical processes would likely be into reversible and irreversible processes. In succeeding lectures, with the aid of the calculus of probability and with the introduction of the hypothesis of elementary disorder, we have seen that all irreversible processes may be considered as reversible elementary processes: in other words, that irreversibility does not depend upon an elementary property of a physical process, but rather depends upon the ensemble of numerous disordered elementary processes of the same kind, each one of which individually is completely reversible, and upon the introduction of the macroscopic method of treatment. From this standpoint one can say quite correctly that in the final analysis all processes in nature are reversible. That there is herein contained no contradiction to the principle regarding the irreversibility of processes expressed in terms of the mean values of elementary processes of macroscopic changes of state, I have demonstrated fully in the third lecture. Perhaps it will be appropriate at this place to interject a more general statement. We are accustomed in physics to seek the explanation of a natural process by the method of division of the process into elements. We regard each complicated process as composed of simple elementary processes, and seek to analyze it through thinking of the whole as the sum of the parts. This method, however, presupposes that through this division the character of the whole is not changed; in somewhat similar manner each measurement of a physical process presupposes that the progress of the phenomena is not influenced by the introduction of the measuring instrument. We have here a case in which that supposition is not warranted, and where a direct conclusion with regard to the parts applied to the whole leads to quite false results. If we divide an irreversible process into its elementary constituents, the disorder and along with it the reversibility's vanishes; an irreversible process must remain beyond the understanding of anyone who relies upon the fundamental law: that all properties of the whole must also be recognizable in the parts. It appears to me as though a similar difficulty presents itself in most of the problems of intellectual life.

Now after all the irreversibility in nature thus appears in a certain sense eliminated, it is an illuminating fact that general elementary dynamics has only to do with reversible processes. Therefore we shall occupy ourselves in what follows with reversible processes exclusively. That which makes this procedure so valuable for the theory is the circumstance that all known reversible processes, be they mechanical, electrodynamical or, thermal, may be brought together under a single principle which answers unambiguously all questions regarding their behavior. This principle is not that of conservation of energy; this holds, it is true, for all these processes, but does not determine unambiguously their behavior; it is the more comprehensive principle of least action.

The principle of least action has grown upon the ground of mechanics where it enjoys equal rank and regard with numerous other principles; the principle of d'Alembert, the principle of virtual displacement, Gauss's principle of least constraint, the Lagrangian Equations of the first and second kind. All these principles are equivalent to one another and therefore at bottom are only different formularizations of the same laws; sometimes one and sometimes another is the most convenient to use. But the principle of least action has the decided advantage over all the other principles mentioned in that it connects together in a single equation the relations between quantities which possess, not only for mechanics, but also for electrodynamics and for thermodynamics, direct significance, namely, the quantities: space, time and potential. This is the reason why one may directly apply the principle of least action to processes other than mechanical, and the result has shown that such applications, as well in electrodynamics as in thermodynamics; lead to the appropriate laws holding in these subjects. Since a representation of a unified system of theoretical physics such as we have here in mind must lay the chief emphasis upon as general an interpretation as possible of physical laws, it is self evident that in our treatment the principle of least action will be called upon to play the principal role. I desire now to show how it is , applied in simple individual cases.

The general formularization of the principle of least action in the interpretation given to it by Helmholz is as follows: among all processes which may carry a certain arbitrarily given physical system subject to given external actions from a given initial position into a given final position in a given time, the process which actually takes place in nature is that which is distinguished by the condition that the integral

$$\int_{t_0}^{t_1} (\delta H + A) dt = 0 , \qquad (57)$$

wherein an arbitrary displacement of the independent coordinates (and velocities) is denoted by the sign δ, and A denotes the infinitely small increase in energy (external work) which the system experiences in the displacement δ. The function H is the kinetic potential. When we speak here of the positions, the coordinates, and the velocities of the configuration, we understand thereby, not only those special ones corresponding to mechanical ideas, but also all the so-called generalized coordinates with the quantities derived therefrom; and these may represent equally well quantities of electricity, volumes, and the like.

In the applications which we shall now make of the principle of least action, we must first decide as to whether the generalized coordinates which determine the state of the system considered are present in finite number or form a continuous infinite manifold. We shall distinguish the examples here considered in accordance with this viewpoint.

Planck's Columbia Lectures

I. *The Position (Configuration) is Determined by a Finite Number of Coordinates.*

In ordinary mechanics this is actually the case in every system of a finite number of material points or rigid bodies among whose coordinates there exist arbitrary fixed equations of condition. If we call the independent coordinates $\varphi_1, \varphi_2, \cdots$, then the external work is:

$$A = \Phi_1 \delta\varphi_1 + \Phi_2 \delta\varphi_2 + \cdots \mp \delta E, \qquad (58)$$

wherein $\Phi_1, \Phi_2, \cdots$ are the "external force components" which correspond to the individual coordinates, and E denotes the energy of the system. Then the principle of least action is expressed by:

$$\int_{t_0}^{t_1} dt \cdot \sum_{1,2,\cdots} \left(\frac{\partial H}{\partial \varphi_1} \delta\varphi_1 + \frac{\partial H}{\partial \dot{\varphi}_1} \delta\dot{\varphi}_1 + \Phi_1 \delta\varphi_1 \right) = 0,$$

From this follow the equations of motion:

$$\Phi_1 - \frac{d}{dt}\left(\frac{\partial H}{\partial \dot{\varphi}_1} \right) + \frac{\partial H}{\partial \varphi_1} = 0, \qquad (59)$$

and so on for all the indices, $1, 2, \cdots$. Through multiplication of the individual equations by $\dot{\varphi}_1, \dot{\varphi}_2, \cdots$. addition and integration with respect to time, there results the equation of conservation of energy, whereby the energy E is given by the expression:

$$E = \sum_{1,2,\cdots} \dot{\varphi}_1 \frac{\partial H}{\partial \dot{\varphi}_1} - H. \qquad (60)$$

In ordinary. mechanics $H = L - U$, if L denote the kinetic and U the potential energy. Since L is a homogeneous function of the second degree with respect to the $\dot{\varphi}$ '*s*, it follows from (60) that:

$$E = 2L - H = L + U.$$

But this expression holds by no means in general.

We pass now to the consideration of the quasi-stationary motion of a system of linear conductors carrying simple closed galvanic currents. The state of the system is given by the position and the velocities of the conductors and by the current densities in each of the same. The coordinates referring to the position of the first conductor may be represented by $\varphi_1, \varphi_1', \varphi_1'', \cdots$ corresponding designations holding for the remaining conductors. We inquire now as to the increase of energy or the external work, A, which corresponds to a virtual displacement of all coordinates. Energy may be conveyed to the system through mechanical actions and through electromagnetic induction as well. The former corresponds to mechanical work, the latter to electromotive work. The former will be of the familiar form:

$$\Phi_1\delta\varphi_1 + \Phi_1'\delta\varphi_1 + \cdots + \Phi_2\delta\varphi_2 + \cdots,$$

If we denote by E_1, E_2, $\cdots$ the electromotive forces which are induced in the .individual conductors through external agencies (e. g., moving magnets which do not belong to the system), then the electromotive work done from outside upon the currents in the conductors of the system is:

$$E_1\delta\,\epsilon_1 + E_2\delta\,\epsilon_2 + \cdots,$$

if $\delta\epsilon_1, \delta\epsilon_2, \cdots$ denote the quantities of electricity which pass through cross sections of the conductors due to infinitely small virtual currents. The finite current densities will then be denoted by $\dot{\epsilon}_1, \dot{\epsilon}_2, \cdots$. The electrical state of the first conductor is thus determined in general by the current density $\dot{\epsilon}_1$, the mechanical state (position and velocity) by the coordinates $\varphi_1, \varphi_1', \varphi_1''$, and the corresponding velocities $\dot{\varphi}_1, \dot{\varphi}_1', \dot{\varphi}_1'', \cdots$. The coordinates $\epsilon_1, \epsilon_2, \cdots$ are so-called " cyclical" coordinates, since the state does not depend upon their momentary values, but only upon their differential quotients with respect to time, just as, for example, the state of a body rotatable about an axis of symmetry depends only upon the angular velocity, and, not upon the angle of rotation. The scheme of notation adopted permits of the direct application of the above formularization of the principle of least action to the case here considered. Thus $H = H_\varphi + H_\epsilon$, where H_φ, the mechanical potential, depends only upon the φ's and $\dot{\varphi}$'s, while the electrokinetic potential H, takes the following form:

$$H_\epsilon = \tfrac{1}{2}L_{11}\,\dot{\epsilon}_1^{\,2} + L_{12}\,\dot{\epsilon}_1\dot{\epsilon}_2 + L_{13}\,\dot{\epsilon}_1\dot{\epsilon}_3 + \cdots \tfrac{1}{2}L_{22}\,\dot{\epsilon}_2^{\,2} + \cdots.$$

The quantities $L_{11}, L_{12}, L_{13}, \cdots L_{22}, \cdots$ the coefficients of self induction and mutual induction depend, however, in a definite manner upon the coordinates of position φ_1, $\varphi_1', \varphi_1'', \cdots \varphi_2, \varphi_2', \varphi_2'', \cdots$.

In accordance with (59), we have for the motion of the first conductor:

$$\Phi_1 - \frac{d}{dt}\left(\frac{dH_\varphi}{\partial \dot{\varphi}_1}\right) + \frac{\partial H_\varphi}{\partial \varphi_1} + \frac{\partial H_\epsilon}{\partial \varphi_1} = 0,$$

with corresponding equations for $\varphi_1, \varphi_1', \varphi_1'', \cdots$, and for the electric current in it:

$$E_1 - \frac{d}{dt}\left(\frac{\partial H_\epsilon}{\partial \dot{\epsilon}_1}\right) = 0.$$

The laws for the mechanical (ponderomotive) actions may be condensed into the statement that, in addition to the ordinary force upon the first conductor expressed by Φ_1, there is a mechanical force

$$\frac{\partial H_\epsilon}{\partial \varphi_1} = \frac{1}{2}\frac{\partial L_{11}}{\partial \varphi_1}\dot{\epsilon}_1^{\,2} + \frac{\partial L_{12}}{\partial \varphi_1}\dot{\epsilon}_1^{\,2}\dot{\epsilon}_2^{\,2} + \frac{\partial L_{13}}{\partial \varphi_1}\dot{\epsilon}_1^{\,2}\dot{\epsilon}_3^{\,2} + \cdots,$$

which is composed of an action of the current upon itself (first term) and of the actions of the remaining currents upon it (following terms).

The laws of electrical action, on the other hand, are expressed by the statement, that to the external electromotive force E_1 in the first conductor there is added the electromotive force

$$-\frac{d}{dt}\left(\frac{dH_\epsilon}{\partial \dot{\epsilon}_1}\right) = -\frac{d}{dt}\left(L_{11}\dot{\epsilon}_1 + L_{12}\dot{\epsilon}_2 + L_{13}\dot{\epsilon}_3 + \cdots\right)$$

which likewise is composed of an action of the current upon itself (self induction) and of the inducing actions of the remaining currents, and that these two forces compensate each other.

The galvanic conductance or the galvanic resistance is not contained in these equations because the corresponding energy, Joule heat, is produced in an irreversible manner, and irreversible processes are not represented by the principle of least action. One can formally include this action, likewise any other irreversible action, in accordance with the procedure of Helmholz, by introducing it as an external force, in the present case as the electromotive force due to the resistance w, which operates to cause a diminution in the energy of the system. For an infinitely small element of time, the amount of this energy change is:

$$-\left(w_1\dot{\epsilon}_1^{\,2} + w_2\dot{\epsilon}_2^{\,2} w_3\dot{\epsilon}_3^{\,2} + \cdots\right)\cdot dt$$
$$= -\left(w_1\dot{\epsilon}_1 d\epsilon_1 + w_2\dot{\epsilon}_2 d\epsilon_2 + \cdots\right)\cdot dt$$

Consequently, since the external work $E_1 d\epsilon_1 + E_2 d\epsilon_2 + \cdots$ now includes the Joule heat, the external force components $E_1, E_2, \cdots$ in the electromotive equations must be increased by the additional terms

$$-w_1\dot{\epsilon}_1, w_2\dot{\epsilon}_2, \cdots.$$

The application of the principle of least action to thermodynamic processes is of special interest, because the importance of the question relating to the fixing of the generalized coordinates, which determine the state of the system, here becomes prominent. From the standpoint of pure thermodynamics, the variables which determine the state of a body can certainly be quite arbitrarily chosen, e.

g., in the case of a gas of invariable. constitution any two of the following quantities may be chosen as independent variables and all others expressed through them: volume V, temperature T, pressure P, energy E, entropy S. In the present case, the matter is quite different. If we inquire, in order to apply the principle of least action, with regard to the energy change or the total work A which will be done upon the gas from without in an infinitely small virtual displacement, it may be written in the form:

$$A = -p \cdot \delta V + \delta S .$$

$T\delta S$ is the heat added from without, $-p\delta V$ the mechanical work furnished from without. In order to bring this into agreement with the general formula for external work (58)

$$A \;=\; \Phi_1 \delta_1 \varphi_1 + \; \Phi_2 \delta_2 \varphi_2$$

it becomes necessary now to choose V and S as the generalized coordinates of state and, therefore, to identify with them the previously employed quantities φ_1 and φ_2. Then $-\ p$ and T are the generalized force components Φ_1 and Φ_2. Now, since in thermodynamics every reversible change of state proceeds with infinite slowness, the velocity components $\dot{V}$ and $\dot{S}$, and in general all differential coefficients with respect to time, are to be placed equal to zero, and the principle of least action (59) reduces to·

$$\Phi \;+\; \frac{\partial H}{\partial \varphi} = 0 \;,$$

and, therefore, in our case:

$$- p + \left(\frac{\partial H}{\partial V} \right)_S = 0 \quad \text{and} \quad T + \left(\frac{\partial H}{\partial S} \right)_V = 0 \ .$$

Further, in accordance with (60)

$$E = -H.$$

Now these equations are actually valid, since they only present other forms of the relation

$$dS = \frac{dE + pdV}{T} \ .$$

The view here presented is fundamentally that which is given in the energetics of Mach, Ostwald, Helm, and Wiedeburg. The generalized coordinates V and S are in this theory the "capacity factors, " $-p$ and T the "intensity factors. "[1] So long as one limits himself to an irreversible process, nothing stands in the way of carrying out this method completely, nor of a generalization to include chemical processes.

In opposition to it there is an essentially different method of regarding thermodynamic processes, which in its complete generality was first introduced into physics by Helmholtz. In accordance with this method, one generalized coordinate is V, and the other is not S, but a certain cyclical coordinate --- we shall denote it, as in

---

1 The breaking up of the energy differentials into two factors by the exponents of energetics is by no means associated with a special property of energy, but is simply an expression for the elementary law that the differential of a function $F(x)$ is equal to the product of the differential dx by the derivative. $\dot{F}(x)$.

the previous example, by ϵ --- which does not appear itself in the expression for the kinetic potential H and only appears through its differential coefficient, $\dot{\epsilon}$; and this differential coefficient is the temperature T. Accordingly, H is dependent only upon V and T. The equation for the total external work, in accordance with (58), is:

$$A = -p\delta V + E\delta\varepsilon,$$

and agreement with thermodynamics is obviously found if we set:

$$E\delta\epsilon = T\delta S, \quad \text{and also:} \quad Ed\epsilon = TdS, \quad Edt = dS.$$

The equations (59) for the principle of least action become:

$$-p = +\left(\frac{\partial H}{\partial V}\right)_T = 0 \quad \text{and} \quad E - \frac{d}{dt}\left(\frac{\partial H}{\partial T}\right)_V = 0,$$

or

$$d\left(\frac{\partial H}{\partial T}\right)_V = Edt = dS,$$

or by integration:

$$\left(\frac{\partial H}{\partial T}\right)_V = S\ ,$$

to an additive constant, which we may set equal to 0. For the energy there results, in accordance with (60)

$$E = \epsilon \frac{\partial H}{\partial \dot{\varepsilon}} - H = T\left(\frac{\partial H}{\partial T}\right)_V - H \ ,$$

and consequently:

$$H = -\ (E\ -\ TS).$$

H is therefore equal to the negative of the function which Helmholz has called the "free energy" of the system, and the above equations are known from thermodynamics.

Furthermore, the method of Helmholz permits of being carried through consistently, and so long as one limits himself to the consideration of reversible processes, it is in general quite impossible to decide in favor of the one method or the other. However, the method of Helmholz possesses a distinct advantage over the other which I desire to emphasize here. It lends itself better to the furtherance of our endeavor toward the unification of the system of physics. In accordance with the purely energetic method, the independent variables V and S have absolutely nothing to do with each other; heat is a form of energy which is distinguished in nature from mechanical energy and which in no way can be referred back to it. In accordance with Helmholz, heat energy is reduced to motion, and this certainly indicates an advance which is to be placed, perhaps, upon exactly the same footing as the advance which is involved in the consideration of light waves as electromagnetic waves.

To be sure, the view of Helmholz is not broad enough to include irreversible processes; with regard to this, as we have earlier stated in detail, the introduction of the calculus of probability is necessary in order to throw light on the question. At the same time, this is also the real reason that the exponents of energetics will have nothing to do with the strict observance of irreversible processes, and they either declare them as doubtful or ignore them completely. In reality, the facts of the case are quite the reverse; irreversible processes are the only processes occurring in nature. Reversible processes form only an ideal abstraction, which is very valuable for the theory, but which is never completely realized in nature.

II. The Generalized Coordinates of State Form a Continuous Manifold.

The laws of infinitely small motions of perfectly elastic bodies furnish us with the simplest example. The coordinates of state are then the displacement components, $\mathfrak{v}_x$, $\mathfrak{v}_y$, $\mathfrak{v}_z$, of a material point from its position of equilibrium $(x, y, . z)$, considered as a function of the coordinates x, y, z. The external work is given by a surface integral:

$$A = \int d\sigma (X_v \delta \mathfrak{v}_z + Y_v \delta \mathfrak{v}_y + Z_v \delta \mathfrak{v}_z)$$

($d\sigma$, surface element; v, inner normal). The kinetic potential is again given by the difference of the kinetic energy L and the potential energy U:

H = L - U.

The kinetic energy is:

$$\mathrm{L} = \int \frac{d\tau k}{2} \left(\dot{\mathfrak{v}}_x^{\,2} + \dot{\mathfrak{v}}_y^{\,2} + \dot{\mathfrak{v}}_z^{\,2} \right),$$

wherein $d\tau$ denotes a volume element, k the volume density. The potential energy U is likewise a space integral of a homogeneous quadratic function f which specifies the potential energy of a volume element. This depends, as is seen from purely geometrical considerations, only upon the 6 “strain coefficients”:

$$\frac{\partial \mathfrak{v}_x}{\partial x} = x_x, \quad \frac{\partial \mathfrak{v}_y}{\partial y} = y_y, \quad \frac{\partial \mathfrak{v}_z}{\partial z} = z_z,$$

$$\frac{\partial \mathfrak{v}_y}{\partial z} + \frac{\partial \mathfrak{v}_z}{\partial y} = y_z = z_y, \quad \frac{\partial \mathfrak{v}_z}{\partial x} + \frac{\partial \mathfrak{v}_x}{\partial z} = z_x = x_z,$$

$$\frac{\partial \mathfrak{v}_x}{\partial y} + \frac{\partial \mathfrak{v}_y}{\partial x} = x_y = y_x.$$

In general, therefore, the function f contains 21 independent constants, which characterize the whole elastic behavior of the substance. For isotropic substances these reduce on grounds of symmetry to 2. Substituting these values in the expression for the principle of least action (57) we obtain:

$$\int dt \left\{ \begin{array}{l} \int d\tau\, k\left(\dot{\mathfrak{v}}_x \delta \dot{\mathfrak{v}}_x + \cdots\right) - \int d\tau \left(\frac{\partial f}{\partial x_x}\delta x_x + \frac{\partial f}{\partial x_y} + \cdots\right) \\ \qquad\qquad\qquad\qquad + \int d\sigma \left(X_y \delta \mathfrak{v}_x + \cdots\right) \end{array} \right\}$$
$$= 0.$$

If we put for brevity:

$$-\frac{\partial f}{\partial x_x} = X_x, \qquad -\frac{\partial f}{\partial y_y} = Y_y, \qquad -\frac{\partial f}{\partial z_z} = Z_z,$$

$$-\frac{\partial f}{\partial y_z} = Y_z = Z_y, -\frac{\partial f}{\partial z_x} = Z_x = X_z, -\frac{\partial f}{\partial x_y} = X_y = Y_x,$$

it turns out, as the result of purely mathematical operations in which the variations $\delta \dot{\mathfrak{v}}_x, \delta \dot{\mathfrak{v}}_x, \cdots$

and likewise the variations $\delta x_x, \delta y_x, \cdots$ are reduced through suitable partial integration with respect to the variations $\delta \mathfrak{v}_x, \delta \mathfrak{v}_x, \cdots$ that the conditions within the body are expressed by:

$$k\ddot{\mathfrak{v}}_x + \frac{\partial X_x}{\partial x} + \frac{\partial X_y}{\partial y} + \frac{\partial X_z}{\partial z} = 0, \cdots$$

and at the surface; by:

$$X_\nu = X_x \cos \nu x + X_y \cos \nu_y + X_z \cos \nu_z, \cdots$$

as is known from the theory of elasticity. The mechanical significance of the quantities X_x, $Y_y, \cdots$ as surface forces follows from the surface conditions.

For the last application of the principle of least action we will take a special case of electrodynamics, namely, electrodynamic processes in a homogeneous isotropic non-conductor at rest, e. g., a vacuum. The treatment is analogous to that carried out in the foregoing example. The only difference lies in the fact that in electrodynamics the dependence of the potential energy U upon the generalized coordinate $\mathfrak{v}$ is somewhat different than in elastic phenomena.

We therefore again put for the external work:

$$A = \int d\sigma \left(X_\nu \delta \mathfrak{v}_x + Y_\nu \delta \mathfrak{v}_y + Z_\nu \delta \mathfrak{v}_z \right), \quad (61)$$

and for the kinetic potential:

$$H = L - U,$$

wherein again:

$$L = \int d\tau \frac{k}{2}\left(\dot{\mathfrak{v}}_x{}^2 + \dot{\mathfrak{v}}_y{}^2 + \dot{\mathfrak{v}}_z{}^2\right) = \int d\tau \frac{k}{2}(\dot{\mathfrak{v}})^2 .$$

On the other hand, we write here:

$$U = \int d\tau \frac{h}{2}(curl\ \mathfrak{v})^2 .$$

Through these assumptions the dynamical equations including the boundary conditions are now completely determined. The principle of least action (57) furnishes:

$$\int dt \left\{ \begin{array}{l} \int d\tau\, k\left(\dot{\mathfrak{v}}_x \delta\, \dot{\mathfrak{v}}_x + \cdots\right) - \int d\tau\, h\left(\mathrm{curl}_x \mathfrak{v} \delta\, \mathrm{curl}_x \mathfrak{v} + \cdots\right) \\ + \int d\sigma\left(X_v \delta\, \mathfrak{v}_x + \cdots\right) \end{array} \right\} = 0.$$

From this follow, in quite an analogous way to that employed above in the theory of elasticity, first, for the interior of the non-conductor:

$$k\dot{\mathfrak{v}}_x = h\left(\frac{\partial\, \mathrm{curl}_y \mathfrak{v}}{\partial z} - \frac{\partial\, \mathrm{curl}_z \mathfrak{v}}{\partial z}\right), \cdots$$

or more briefly

$$k\ddot{\mathfrak{v}} = -\, h\, \mathrm{curl}\, \mathrm{curl}\, \mathfrak{v}, \qquad (62)$$

and secondly, for the surface:

$$X_v = h\left(\mathrm{curl}_z \mathfrak{v} \cdot \cos vy - \mathrm{curl}_y \mathfrak{v} \cdot \cos \ldots vz\right), \cdots . \quad (63)$$

These equations are identical with the known electrodynamical equations, if we identify L with the electric, and U with the magnetic energy (or conversely). If we put

$$L = \frac{1}{8\pi}\int d\tau \cdot \epsilon \mathfrak{E}^2 \quad \text{and} \quad U = \frac{1}{8\pi}\int d\tau \cdot \mu\, \mathfrak{H}^2,$$

($\mathfrak{E}$ $\mathfrak{H}$ the field strengths, ϵ, the dielectric constant, μ, the permeability) and compare these values with the above expressions for L and U we may write:

$$\dot{\mathfrak{v}} = -\mathfrak{E}\cdot\sqrt{\frac{\epsilon}{4\pi k}}, \quad \text{curl}\, \mathfrak{H}\cdot\sqrt{\frac{\mu}{4\pi h}}. \quad (64)$$

It follows then, by elimination of $\mathfrak{v}$, that:

$$\dot{\mathfrak{H}} = -\sqrt{\frac{\epsilon h}{\mu k}}\, \text{curl}\, \mathfrak{E},$$

and further, by substitution of $\dot{\mathfrak{v}}$ and curl $\mathfrak{v}$ in equation (62) found above for the interior of the non-conductor, that:

$$\dot{\mathfrak{E}} = -\sqrt{\frac{\mu h}{\epsilon k}}\, \text{curl}\, \mathfrak{H}.$$

From either of these two equations it follows that: Comparison with the known electrodynamical equations expressed in Gaussian units:

$$\mu\dot{\mathfrak{H}} = -c\, \text{curl}\, \mathfrak{E}, \quad \epsilon\dot{\mathfrak{E}} = -c\, \text{curl}\, \mathfrak{H}$$

(c, velocity of light in vacuum) results in a complete agreement, if we put:

$$\frac{c}{\mu} = -\sqrt{\frac{\epsilon h}{\mu k}} \text{ and } \frac{c}{\epsilon} = -\sqrt{\frac{\mu h}{\epsilon k}},$$

$$\frac{h}{k} = \frac{c^2}{\epsilon\mu},$$

for the square of the velocity of propagation c.

We obtain from. (61) for the energy entering the system from without:

$$dt \cdot \int d\sigma \left(X_\nu \dot{\mathfrak{v}}_x + Y_\nu \dot{\mathfrak{v}}_y + Z_\nu \dot{\mathfrak{v}}_z \right),$$

or, taking account of the surface equation (63):

$$dt \cdot \int d\sigma\, h \left\{ \left(\text{curl}_z \mathfrak{v} \cos \nu y - \text{curl}_y \mathfrak{v} \cos \nu z \right) \dot{\mathfrak{v}}_x + \cdots \right\},$$

an expression which, upon substitution of the values of $\dot{\mathfrak{v}}$ and curl $\mathfrak{v}$ from (64), turns out to be identical with the Poynting energy current.

We have thus by an application of the principle of least action with a suitably chosen expression for the kinetic potential H arrived at the known Maxwellian field equations.

Are, then, the electromagnetic processes thus referred back to mechanical processes? By no means; for

the vector $\mathfrak{v}$ employed here is certainly not a mechanical quantity. It is moreover not possible in general to interpret $\mathfrak{v}$ as a mechanical quantity, for instance, $\mathfrak{v}$ as a displacement, $\dot{\mathfrak{v}}$ as a velocity, curl $\mathfrak{v}$ as a rotation. Thus, e. g., in an electrostatic field $\dot{\mathfrak{v}}$ is constant. Therefore, $\mathfrak{v}$ increases with the time beyond
all limits, and curl $\mathfrak{v}$ can no longer signify a rotation.[2] While from these considerations the possibility of a mechanical explanation of electrical phenomena is not proved, it does appear, on the other hand, to be undoubtedly true that the significance of the principle of least action may be essentially extended beyond ordinary mechanics and that this principle can therefore also be utilized as the foundation for general dynamics, since it governs all known reversible processes.

---

2 With regard to the impossibility of interpreting electrodynamic processes in terms of the motions of a continuous medium, cf. particularly, H. Witte: "Über den gegenwärtigen Stand der Frage nach einer mechanischen Erklärung der elektrischen Erscheinungen " Berlin, 1906 (E. Ebering).

Planck's Columbia Lectures

Summary of the Seventh Lecture

The characteristics of reversible and irreversible processes were reviewed at the beginning of this lecture. According to Planck, irreversibility depends upon the "ensemble of numerous disordered elementary processes of the same kind, each one of which is completely reversible". In an irreversible process, the average or mean value of the elementary process represents the state of the element within a macroscopic region. Irreversible processes are therefore probabilistic, and the entropy equation is a measure of the disorder of a system in terms of the number of energy states and the number of molecules. In probability theory, the properties of the unknowns occur in patterns that can be characterized, resulting in various types of general equations. For instance, the probability equations for the flipping of a coin has only two possible outcomes or states, and its probability distribution is different than the one developed here. The probability equation for W, is basic to irreversible processes of a large number of molecules, and Planck was able to derive the equation for the probability distribution in its exact form. Once the probability distribution was determined, the entropy equation was then complete, leading to the energy state equation and the radiation equation.

It may seem somewhat confusing that all elementary systems are reversible, while a process that contains a large number of reversible systems can be irreversible. One way to look at it is that the energy can be restored to a reversible system, but that reversible systems generally lose a small amount of energy outside of the system through the process of radiation, and this energy cannot be restored. The use of this approach has another advantage. One can question whether or not the actions of a system consisting of a large number of elements can be exactly determined from the actions of the elements. If we knew the response functions of the molecules, their initial states, and the driving functions affecting each molecule at that point in time, it is conceivable that the entire system might be characterized, and the probability functions would not come into play. However, for a molecular mass, there are a huge numbers of molecules and a great number of energy states to be considered, and this would be an enormous task, even if it were possible to determine all of the initial conditions for all of the molecules and their driving functions. Therefore, it is much more practical to consider it as an irreversible system and utilize probabilistic determinations as Planck has done, rather than attempting to analyze it as a reversible system consisting of an enormous number of energy states for a huge number of molecules whose initial states are undetermined.

Planck's energy method of analysis has proved to be exceptionally accurate in analyzing the average energy states of irreversible systems. In this lecture, however, Planck points out that the principle of conservation of energy does not unambiguously determine the behavior of these processes. He stated that the Principle of Least Action is more comprehensive in making these determinations. Presumably, this principle goes beyond the *unified field theory* that Einstein sought, since it applies, not only to fields, but to mechanics, thermodynamics and electromagnetics, each of which is a function of *space*, *time*, *potential* and *force*, and therefore potentially leading to a "unified system of theoretical physics". The material that he presented in this lecture is an important initial step in that process.

As stated earlier, the original formats have been retained for the lectures. The symbols that Planck uses may represent different functions in the various scenarios. In some cases, they are also different from today's common terminology, and these differences will be elucidated as applicable.

The Principle of Least Action, formulated by Helmholtz, is expressed by equation (57)

$$\int_{t_0}^{t_1} (\delta H + A)dt = 0$$

where δ denotes the displacement of independent coordinates or velocities, A denotes the small increase in energy from external work that the system experiences, and H is the kinetic potential. A system consists of a number of elements and coordinates that may be either finite in number or form a continuous manifold, and Planck uses two sets of examples for these two types of systems.

I. The Position (Configuration) is Determined by a Finite Number of Coordinates.]

The first set of examples are for a finite number of coordinates, which is the case for ordinary mechanics. In the first example, he considers a mechanical system of rigid bodies. For *n* independent coordinates, φ_n, the external work is

$$A = \sum_n \Phi_n \delta\varphi_n = \delta E.$$

The Φ_n are the external force components and *E* is the energy of the system. This is equation (58), for the energy increase of the system, δE, is written above in the form of a summation, which is a more compact was to write the equation.

The differential of the kinetic potential *H* is

$$\delta H = \sum_n \left(\frac{\partial H}{\delta\varphi_n} \delta\varphi_n + \frac{\partial H}{\delta\dot{\varphi}_n} \delta\dot{\varphi}_n \right).$$

Note that there are two components within this equation; one for the position and one for the velocity of each independent coordinate. Substituting into the integral equation,

$$\int_{t_0}^{t_1} dt \cdot \sum_n \left(\frac{\partial H}{\delta\varphi_n} \delta\varphi_n + \frac{\partial H}{\delta\dot{\varphi}_n} \delta\dot{\varphi}_n + \Phi_n \delta\varphi_n \right) = 0.$$

Equating term-by-term,

$$\Phi_n - \frac{d}{dt}\left(\frac{\partial H}{\partial \dot{\varphi}_n}\right) + \frac{\partial H}{\partial \phi_n} = 0,$$

which represents the Lagrangian equations of motion (59) for each of the *n* coordinate dimensions. This equation is valid for very small values of $\delta\varphi$ and $\delta\dot{\varphi}$ within the integral. The left hand term Φ_n correlates to the A term (the energy differential) within the integral of equation (57). Planck reduced this equation by first multiplying all terms by $\dot{\varphi}$ and then solving for the energy,

$$E = \sum_n \dot{\varphi}_n \frac{\partial H}{\partial \dot{\varphi}_n} - H,$$

which is equation (60). Note that there are only two variables , $\dot{\varphi}_n$ and *H*, for each term within the summation. Thus the equations of motion and the expression for the system energy follow from the Principle of Least Action. In the applications that follow, Planck applies external work equation (58), the equations of motion (59) and the system energy equation (60) in analyzing the actions of systems.

For mechanical systems,

$$H = L - U$$

where *L* is the kinetic energy and *U* is the potential energy. The kinetic energy is a homogeneous function of second degree (in today's world we would write the variable as *v* for velocity, rather than as $\dot{\varphi}$), the resulting derivative in the summation term of equation (60) has a coefficient of 2, and

$$E = 2L - H = L + U.$$

The total energy of the mechanical system is the sum of the kinetic and potential energies, as one would expect.

The next example describes a system that consists of a system of linear conductors carrying electric currents. Energy can be introduced in this system through both mechanical actions and electromagnetic action. The equation for mechanical work is introduced in the same manner as was done above,

$$A = \delta E = \sum \Phi_n \delta\varphi_n .$$

Electromotive forces, E_n, are induced in the conductors by electromagnetic fields, such as in the case for moving magnets located outside the system. The work done on the system is then

$$W = \sum_n E_n \delta\epsilon_n .$$

Note the change in form of the variables in this equation as compared those used for mechanical systems. The $\delta\epsilon_n$ terms represent the amount of charge passing through the conductors (in contemporary terminology, we would write $\delta\epsilon_n$ as q_n, and the current density, $\dot{\epsilon}_n$, as J_n). Planck called the charge coordinates ϵ_n cyclical, since the state of a lossless inductor depends only on the time derivatives rather than the instantaneous values. He cited a similar example for the case of mechanical body rotating about an axis, whose state depends upon its rotational velocity.

Planck defined the movement of charges in terms of their *velocities*, rather than the contemporary definition of electrical current that portrays it as simply the number of charges per unit time that are moving through a conductor. This more thorough definition allows the analysis to extend over the regions of mechanical and electrical forces. The kinetic potential consists of two components

$$H = H_{\phi} + H_{\epsilon},$$

where H_{ϕ} is the magnetic potential and H_{ϵ} is the electrokinetic potential. These potentials are defined in terms of equation (57), which defines the Principle of Least Action.

The electrokinetic potential is

$$H_{\epsilon} = \sum_{1,n} \left(\frac{1}{2} L_{11} \dot{\epsilon}^{2} + L_{1n} \dot{\epsilon}_1 \dot{\epsilon}_n \right),$$

The L_{11} term is the self inductance, and the other L_{1n} terms represent the mutual inductances. The n coordinate positions and velocities are all measured with respect to the position of the first element.

Substituting into equation (59) for the motion of the first conductor,

$$\Phi_1 - \frac{d}{dt}\left(\frac{dH_{\phi}}{d\dot{\varphi}_1} \right) + \frac{\partial H_{\phi}}{\partial \varphi_1} + \frac{\partial H_{\epsilon}}{\partial \varphi_1} = 0,$$

where Φ_1 is the external mechanical force exerted on the first conductor. For the electric current in the first conductor,

$$E_1 = \frac{d}{dt}\left(\frac{\partial H_\epsilon}{\partial \dot{\epsilon}_1}\right).$$

The relative positions and motions of the other inductive elements with respect to the first element yield a set of similar equations. The force exerted on the first conductor by the external force Φ_1 is determined in this manner.

In addition to the force exerted on the first conductor by Φ_1, there are other mechanical forces produced by the electrical currents,

$$\frac{\partial H}{\partial \varphi_1} = \frac{1}{2}\frac{\partial L_{11}}{\partial \varphi_1}\dot{\epsilon}_1^{\,2} + \sum_2^n \frac{\partial L_{1,n}}{\partial \varphi_1}\dot{\epsilon}_1\dot{\epsilon}_n .$$

The first term to the right of the equal sign is the force acting on the first element that is due to its self inductance, while the forces within the summation are due to mutual inductances existing between the first element and the other elements. This equation was obtained by partial differentiation of the electrokinetic potential equation shown above. In addition to the external electromotive force E_1, there is another electromotive force, produced by induction, which compensates for the electromagnetic coupled force,

$$-\frac{d}{dt}\left(\frac{dH_\epsilon}{\partial \dot{\epsilon}_1}\right) = -\frac{d}{dt}\left(\sum_n L_{1n}\dot{\epsilon}_n\right).$$

In this step, Planck differentiated with respect to current and time, rather than to the mechanical coordinate variables. The first term within the summation is the inductive voltage drop across the first inductor, L_{11}, produced by the current through it, and the other terms are the voltages produced in the first inductor by the currents in the other inductors.

All of the inductors have resistance, and the current passing them produces "Joule heat". This is an irreversible process that is not represented in the Principle of Least Action equation. Helmholtz was able to accommodate this difficulty by representing it as an external force that produces an energy loss in the system due to the resistance (the variable *w* for resistance). The resulting energy change is

$$dE = -\left(\sum_n w_n \dot{\epsilon}_n^{\,2}\right) \cdot dt = -\sum_n w_n \dot{\epsilon}_n d\epsilon_n ,$$

The external force components Φ_n must include these additional $-w_n \dot{\epsilon}_n$ (electromotive force) terms.

In the next example, Planck applied the Principle of Least Action to thermodynamic processes. In the prior thermodynamic analysis of energy states, the exact determination of the space coordinates was not necessary, since the variables (*V*,*T*, *P*, *S*) were not represented directly in terms of position or velocity. The total work exerted on a gas by an infinitesimal displacement is, by equation (58),

$$A = -p \cdot \delta V + T \cdot \delta S = \Phi_1 \delta\varphi_1 + \Phi_2 \delta\varphi_2 .$$

In this equation, the first term represents the heat absorbed by the system, and the second term is the mechanical work provided by an external source. The state coordinates are defined as *V* and *S*, and their time derivatives are zero for every reversible change of state. Applying the equation of motion (59),

$$-p+\left(\frac{\partial H}{\partial V}\right)_S=0 \quad \text{and} \quad T+\left(\frac{\partial H}{\partial S}\right)_V=0,$$

which is a form of the earlier relation

$$dS=\frac{dE+pdV}{T}$$

discussed in the second lecture on thermodynamic processes.

These illustrations show that Helmholtz's Principle of Least Action works equally as well as the earlier methods for reversible processes. For irreversible processes, it was shown that additional steps are required in the analysis, and the calculus of probability comes into play. Although this is a difficulty that must be overcome, there is a very important additional factor in this method of analysis that applies to reversible systems. A special term appears when taking the derivative of the energy equation (60),

$$E=\dot{\epsilon}\frac{\partial H}{\partial \dot{\epsilon}}-H.$$

Solving this equation, the kinetic potential is

$$H=-(E-TS).$$

The above equations that were applied to the Principle of Least Action are from thermodynamics, and Helmholtz called the following term,

$$(E - TS).$$

The first term, E, is the total energy, and the other term is the heat energy. The remaining energy is the "free energy" of the system.

As Planck viewed it, heat energy is reduced to motion, and yet it is different from mechanical energy "and which in no way can be referred back to it". The wording of this statement is a bit confusing, since mechanical energy can be related to heat energy by the exchange of energy occurring in the process. This may have been a difficulty in translation, and perhaps a better way to say it is that the process of radiation of energy is quite different from the process that produces a loss of mechanical energy in heat exchange. Planck ranked the importance of reducing heat energy to motion as being on a par with the consideration of light waves as electromagnetic waves.

II. The Generalized Coordinates of State Form a Continuous Manifold.

There are two examples in this section. The first example is for a system of perfectly elastic bodies in which the "generalized coordinates of state form a continuous manifold". There are six strain coefficients, and the potential energy *U* is a space integral of a quadratic function *f* that contains 21 independent constants, so this will be a bit more complex.

The result of applying the Principle of Least Action, produces two expressions that define both the conditions within the body and at the surface. The coordinates of state are the displacement components, $\mathfrak{v}_x$, $\mathfrak{v}_y$, $\mathfrak{v}_z$, of a material point from its position of equilibrium (x_0, y_0, z_0) as a function of the coordinate variables *x*, *y*, *z*. The external work is given by a surface integral:

$$A = \int d\sigma (X_v \delta \mathfrak{v}_z + Y_v \delta \mathfrak{v}_y + Z_v \delta \mathfrak{v}_z)$$

($d\sigma$, surface element; *v*, inner normal). The kinetic potential is the difference of the kinetic energy *L* and the potential energy *U*:

H = L - U.

The second expression is for the kinetic energy:

$$\int \mathrm{L} = \int \frac{d\tau k}{2} \left(\dot{\mathfrak{v}}_x{}^2 + \dot{\mathfrak{v}}_y{}^2 + \dot{\mathfrak{v}}_z{}^2 \right),$$

wherein $d\tau$ denotes a volume element and k is the

volume density. The potential energy U is also a space integral of a homogeneous quadratic function f which specifies the potential energy of a volume element. This depends (geometrically) upon the six "strain coefficients":

$$\frac{\partial \mathfrak{v}_x}{\partial x} = x_x, \quad \frac{\partial \mathfrak{v}_y}{\partial y} = y_y, \quad \frac{\partial \mathfrak{v}_z}{\partial z} = z_z,$$

$$\frac{\partial \mathfrak{v}_y}{\partial z} + \frac{\partial \mathfrak{v}_z}{\partial y} = y_z = z_y, \quad \frac{\partial \mathfrak{v}_z}{\partial x} + \frac{\partial \mathfrak{v}_x}{\partial z} = z_x = x_z,$$

$$\frac{\partial \mathfrak{v}_x}{\partial y} + \frac{\partial \mathfrak{v}_y}{\partial x} = x_y = y_x.$$

In general, the function f contains 21 independent constants that characterize the whole elastic behavior of the substance. For isotropic substances these reduce (on grounds of symmetry) to just two. Substituting the above values into the expression for the principle of least action (57):

$$\int dt \left\{ \int d\tau\, k\left(\dot{\mathfrak{v}}_x \delta \dot{\mathfrak{v}}_x + \cdots\right) - \int d\tau \left(\frac{\partial f}{\partial x_x} \delta\, x_x + \frac{\partial f}{\partial x_y} + \cdots \right) + \int d\sigma \left(X_y \delta\, \mathfrak{v}_x + \cdots \right) \right\} = 0.$$

Using the following definitions:

$$-\frac{\partial f}{\partial x_x} = X_x, \qquad -\frac{\partial f}{\partial y_y} = Y_y, \qquad -\frac{\partial f}{\partial z_z} = Z_z,$$

$$-\frac{\partial f}{\partial y_z} = Y_z = Z_y, -\frac{\partial f}{\partial z_x} = Z_x = X_z, -\frac{\partial f}{\partial x_y} = X_y = Y_x,$$

the variations $\delta\dot{\mathfrak{v}}_x, \delta\dot{\mathfrak{v}}_x, \cdots$ and $\delta x_x, \delta y_x, \cdots$, through suitable partial integration with respect to the variations $\delta\mathfrak{v}_x, \delta\mathfrak{v}_x, \cdots$, result in the following conditions within the body:

$$k\ddot{\mathfrak{v}}_x + \frac{\partial X_x}{\partial x} + \frac{\partial X_y}{\partial y} + \frac{\partial X_z}{\partial z} = 0, \cdots$$

and at the surface by:

$$X_\nu = X_x \cos \nu x + X_y \cos \nu_y + X_z \cos \nu_z, \cdots,$$

which is known from the theory of elasticity. The quantities X_x, Y_y, $\cdots$ are mechanical surface forces that depend upon the surface conditions.

The second example is the last application of the Principle of Least Action. It is a special case of electrodynamics, which is an electrodynamic process within a homogeneous isotropic non-conductor at rest (a vacuum). The treatment is similar to the foregoing example. The only difference is that, in electrodynamics, the dependence of the potential energy U upon the generalized coordinate $\mathfrak{v}$ is somewhat different than in elastic phenomena.

The external work is expressed as:

$$A = \int d\sigma \left(X_\nu \delta\mathfrak{v}_x + Y_\nu \delta\mathfrak{v}_y + Z_\nu \delta\mathfrak{v}_z \right), \quad (61)$$

and the kinetic potential is:

$$H = L - U,$$

and therefore,

$$L = \int d\tau \frac{k}{2}\left(\dot{\mathfrak{v}}_x{}^2 + \dot{\mathfrak{v}}_y{}^2 + \dot{\mathfrak{v}}_z{}^2\right) = \int d\tau \frac{k}{2}(\dot{\mathfrak{v}})^2$$

The energy is expressed by an integral equation:

$$U = \int d\tau \frac{h}{2}(\text{curl } \mathfrak{v})^2 ,$$

and the dynamical equations, including the boundary conditions, are now completely determined. Applying the Principle of Least Action (57):

$$\int dt \left\{ \begin{aligned} &\int d\tau\, k\left(\dot{\mathfrak{v}}_x \delta\, \dot{\mathfrak{v}}_x + \cdots\right) - \int d\tau\, h\left(\text{curl}_x \mathfrak{v} \delta \,\text{curl}_x \mathfrak{v} + \cdots\right) \\ &+ \int d\sigma \left(X_\nu \delta\, \mathfrak{v}_x + \cdots \right) \end{aligned} \right\} = 0.$$

Using the methods employed above for the theory of elasticity, within the <u>interior</u> of the non-conductor:

$$k\ddot{\mathfrak{v}}_x = h\left(\frac{\partial \,\text{curl}_y \mathfrak{v}}{\partial z} - \frac{\partial \,\text{curl}_z \mathfrak{v}}{\partial z} \right), \cdots$$

which can be written more simply as:

$$k\ddot{\mathfrak{v}} = -h \,\text{curl curl}\, \mathfrak{v}, \qquad (62)$$

and at <u>the surface</u>:

$$X_\nu = h\left(\text{curl}_z \mathfrak{v} \cdot \cos \nu y - \text{curl}_y \mathfrak{v} \cdot \cos \ldots \nu z\right), \cdots \quad (63)$$

If we identify L with the electric, and U with the magnetic energy,these equations are identical with the known electrodynamical equations. If we put

$$L = \frac{1}{8\pi} \int d\tau \cdot \epsilon \mathfrak{E}^2 \quad \text{and} \quad U = \frac{1}{8\pi} \int d\tau \cdot \mu \mathfrak{H}^2,$$

($\mathfrak{E}$ and $\mathfrak{H}$ are the field strengths, ϵ is the dielectric constant, and μ i s the permeability for this expression) and compare these values with the above expressions for L and U:

$$\dot{\mathfrak{v}} = -\mathfrak{E} \cdot \sqrt{\frac{\epsilon}{4\pi k}}, \quad \text{curl}\, \mathfrak{H} \cdot \sqrt{\frac{\mu}{4\pi h}}. \quad (64)$$

It follows then, by elimination of $\mathfrak{v}$, that:

$$\dot{\mathfrak{H}} = -\sqrt{\frac{\epsilon h}{\mu k}}\, \text{curl}\, \mathfrak{E}.$$

By substitution of $\dot{\mathfrak{v}}$ and curl $\mathfrak{v}$ into equation (62) (found above for the interior of the non-conductor):

$$\dot{\mathfrak{E}} = -\sqrt{\frac{\mu h}{\epsilon k}}\, \text{curl}\, \mathfrak{H}.$$

With the known electrodynamical equations expressed in Gaussian units:

$$\mu \dot{\mathfrak{H}} = -c\, \text{curl}\, \mathfrak{E}, \quad \epsilon \dot{\mathfrak{E}} = -c\, \text{curl}\, \mathfrak{H}$$

(c is the velocity of light in a vacuum) results in a complete agreement, if we put:

$$\frac{c}{\mu} = -\sqrt{\frac{\epsilon h}{\mu k}} \text{ and } \frac{c}{\epsilon} = -\sqrt{\frac{\mu h}{\epsilon k}}.$$

From either of these two equations it follows that:

$$\frac{h}{k} = \frac{c^2}{\epsilon\mu},$$

the square of the velocity of propagation.

Utilizing equation (61), we obtain the energy entering the system from without:

$$dt \cdot \int d\sigma \left(X_v \dot{\mathfrak{v}}_x + Y_v \dot{\mathfrak{v}}_y + Z_v \dot{\mathfrak{v}}_z \right),$$

or by the surface equation (63):

$$dt \cdot \int d\sigma \, h \left\{ \left(\text{curl}_z \mathfrak{v} \cos vy - \text{curl}_y \mathfrak{v} \cos vz \right) \dot{\mathfrak{v}}_x + \cdots \right\},$$

from which the substitution of the values of $\dot{\mathfrak{v}}$ and curl $\mathfrak{v}$ from (64) turns out to be identical with the Poynting energy current.

Thus, by an application of the principle of least action with a suitably chosen expression for the kinetic potential H, Planck arrived at the Maxwellian field equations.

Recounting the steps in this last example, the Principle of Least Action was applied to the electrodynamics of an isotropic medium, which in this case is a vacuum. Planck then applied a method that is similar to that used for the theory of elasticity. After deriving the proper expression for the kinetic potential *H*, he derived Maxwell's volumetric field equations and the Poynting energy, which is a surface function.

In electromagnetic analysis, the displacement components are space functions, as was the case for mechanical systems. However, Planck pointed out that it is not generally possible to interpret the displacement variable $\mathfrak{v}$ as a mechanical quantity for electromagnetic systems. The reason for this difference is that the rotation of the field is curl $\mathfrak{v}$, and therefore the velocity $\dot{\mathfrak{v}}$ is constant for an electrostatic field. With a constant velocity, the displacement increases constantly with time, and therefore "curl $\mathfrak{v}$ can no longer signify a rotation". There is a problem with the interpretation of the spatial characteristics of electrical phenomena as mechanical forms (electrical phenomena are "perfectly elastic"), and Planck cited a reference regarding this difficulty. However, he maintained that the Principle of Least Action can be extended beyond ordinary mechanics, and that this principle can be utilized as the foundation for general dynamics, since "it governs all known reversible processes".

The material in this chapter was a precursor of electrical circuit theory and finite element analysis of mechanical systems. In fact, it extends beyond these modern methods, in certain respects, in that all possible conditions are considered. Also, Planck defined the variables of electrical systems in terms of electric charge, rather than current, and included the space variables (which, for instance, is not present in the definition of electrical current).

In the next chapter on the principle of relativity, there is a basic hypothesis which establishes the dependence of the kinetic potential upon the generalized coordinates and their derivatives (velocities). In the final chapter, I will present an alternate method of analysis that allows a unique definition of curl $\mathfrak{v}$ that leads to some interesting conclusions.

Planck's Columbia Lectures

Chapter 8

Introduction to the Eighth Lecture

In this final lecture, Planck discusses the theory of relativity. In these few pages, he presents a fairly thorough description of how relativity theory relates to the laws of physics and real world measurements. The last part of this lecture contains some interesting observations.

As you read Planck's description of the General Theory of Relativity, carefully note the assumptions upon which the theory is based, as the theory depends upon them.

The next chapter of this book contains a summary of all of Planck's lectures, and the final chapter contains a discussion of some of the developments that have occurred in the better part of a century that have passed since he delivered these lectures and how his theories fit in with them.

EIGHTH LECTURE

GENERAL DYNAMICS. PRINCIPLE OF RELATIVITY.

In the lecture of yesterday we saw, by means of examples, that all continuous reversible processes of nature may be represented as consequences of the principle of least action, and that the whole course of such a process is uniquely determined as soon as we know, besides the actions which are exerted upon the system from without, the kinetic potential H as a function of the generalized coordinates and their differential coefficients with respect to time. The determination of this function remains then as a special problem, and we recognize here a rich field for further theories and hypotheses. It is my purpose to discuss with you today an hypothesis which represents a magnificent attempt to establish quite generally the dependency of the kinetic potential H upon the velocities, and which is commonly designated as the principle of relativity. The gist of this principle is: it is in no wise possible to detect the motion of a body relative to empty space; in fact, there is absolutely no physical sense in speaking of such a motion. If, therefore, two observers move with uniform but different velocities, then each of the two with exactly the same right may assert that with respect to empty space he is at rest, and there are no physical methods of measurement enabling us to decide in favor of the one or the other. The principle of relativity in its generalized form is a very recent development. The preparatory steps were taken by H. A. Lorentz, it was first generally formulated by A. Einstein, and was developed into a finished mathematical system by H. Minkowski. However, traces of it extend quite far back into the past, and therefore it seems desirable first to say something concerning the history of its development.

The principle of relativity has been recognized in mechanics since the time of Galileo and Newton. It is contained in the form of the simple equations of motion of a material point, since these contain only the acceleration and not the velocity of the point. If, therefore, we refer the motion of the point, first to the coordinates $x, y, z,$ and again to the coordinates x', y', z' of a second system, whose axes are directed parallel to the first and which moves with the velocity v in the direction of the positive x-axis:

$$x' = x - vt, \quad y' = y, \quad z' = z, \qquad (65)$$

and the form of the equations of motion is not changed in the slightest. Nothing short of the assumption of the general validity of the relativity principle in mechanics can justify the inclusion by physics of the Copernican cosmical system, since through it the independence of all processes upon the earth of the progressive motion of the earth is secured. If one were obliged to take account of this motion, I should have, e. g., to admit that the piece of chalk in my hand possesses an enormous kinetic energy, corresponding to a velocity of something like 30 kilometers per second.

It was without doubt his conviction of the absolute validity of the principle of relativity which guided Heinrich Hertz in the establishment of his fundamental equations for the electrodynamics of moving bodies. The electrodynamics of Hertz is, in fact, wholly built upon the principle of relativity. It recognizes no absolute motion with regard to empty space. It speaks only of motions of material bodies relative to one another. In accordance with the theory of Hertz, all electrodynamic processes occur in material bodies; if these move, then the electrodynamic processes occurring therein move with them. To speak of an independent state of motion of a medium outside of material bodies, such as the ether, has just as little sense in the theory of Hertz as in the modern theory of relativity.

But the theory of Hertz has led to various contradictions with experience. I will refer here to the most important of these. Fizeau brought (1851) into parallelism a bundle of rays originating in a light source L by means of a lens and then brought it to a focus by means of a second lens upon a screen S
(Fig. 2).

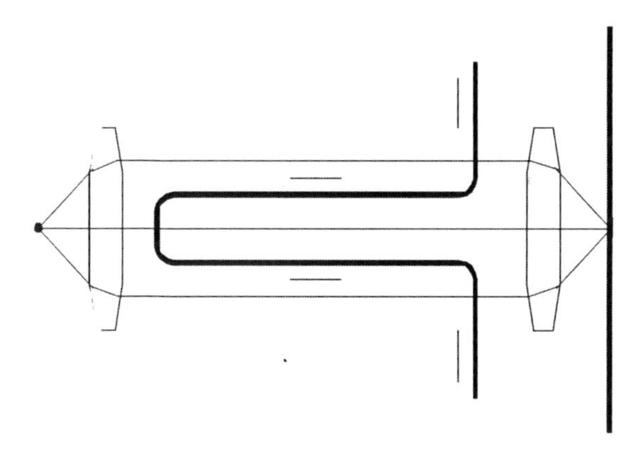

Fig. 2.

In the path of the parallel light rays between the two lenses he placed a tube system of such sort that a transparent liquid could be passed through it, and in such manner that in one half (the upper) the light rays would pass in the direction of flow of the liquid while in the other half (the lower), the rays would pass in the opposite direction.

If now a liquid or a gas flow through the tube system with the velocity v, then, in accordance with the theory of Hertz, since light must be a process in the substance, the light waves must be transported with the velocity of the liquid. The velocity of light relative to L and S is, therefore, in the upper part $q_0 + v$, and the lower part $q_0 - v$, if q_0 denote the velocity of light relative to the liquid. The difference of these two velocities, $2v$, should be observable at S through corresponding interference of the lower and the upper light rays, and quite independently of the nature of the flowing substance. Experiment did not confirm this conclusion. Moreover, it showed in gases generally no trace of the expected action; i.e., light is propagated in a flowing gas in the same manner as in a gas at rest. On the other hand, in the case of liquids an effect was certainly indicated, but notably smaller in amount than that demanded by the theory of Hertz. Instead of the expected velocity difference $2v$, the difference $2v(1 - 1/n^2)$ only was observed, where n is the refractive index of the liquid. The factor $(1 - 1/n^2)$ is called the Fresnel coefficient. There is contained (for $n = 1$) in this expression the result obtained in the case of gases.

It follows from the experiment of Fizeau that, as regards electrodynamic processes in a gas, the motion of the gas is practically immaterial. If, therefore, one holds that electrodynamic processes require for their propagation a substantial carrier, a special medium, then it must be concluded that this medium, the ether, remains at rest when the gas moves in an arbitrary manner. This interpretation forms the basis of the electrodynamics of Lorentz, involving an absolutely quiescent ether. In accordance with this theory, electrodynamic phenomena have only indirectly to do with the motion of matter. Primarily all electrodynamical actions are propagated in ether at rest.

Matter influences the propagation only in a secondary way, so far as it is the cause of exciting in greater or less degree resonant vibrations in its smallest parts by means of the electrodynamic waves passing through it. Now, since the refractive properties of substances are also influenced through the resonant vibrations of its smallest particles, there results from this theory a definite connection between the refractive index and the coefficient of Fresnel, and this connection is, as calculation shows, exactly that demanded by measurements. So far, therefore, the theory of Lorentz is confirmed through experience, and the principle of relativity is divested of its general significance.

The principle of relativity was immediately confronted by a new difficulty. The theory of a quiescent ether admits the idea of an absolute velocity of a body, namely the velocity relative to the ether. Therefore, in accordance with this theory, of two observers A and B who are in empty space and who move relatively to each other with the uniform velocity v, it would be at best possible for only one rightly to assert that he is at rest relative to the ether. If we assume, e. g., that at the moment at which the two observers meet an instantaneous optical signal, a flash, is made by each, then an infinitely thin spherical wave spreads out from the place of its origin in all directions through empty space. If, therefore, the observer A remain at the center of the sphere, the observer B will not remain at the center and, as judged by the observer B, the light in his own direction of motion must travel (with the velocity $c - v$) more slowly than in the opposite direction (with the velocity $c + v$), or than in a perpendicular direction (with the velocity $\sqrt{c^2 - v^2}$) (cf. Fig. 3).

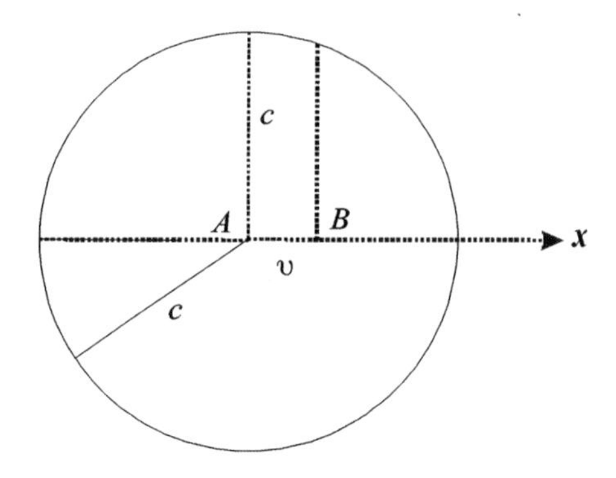

Fig. 3.

Under suitable conditions the observer B should be able to detect and measure this sort of effect.

This elementary consideration led to the celebrated attempt of Michelson to measure the motion of the earth relative to the ether. A parallel beam of rays proceeding from L (Fig. 4) falls upon a transparent plane parallel plate P inclined at 45°, by which it is in .part transmitted and in part reflected. The transmitted and reflected beams are brought into interference by reflection from suitable metallic mirrors S_1 and S_2, which are removed by the

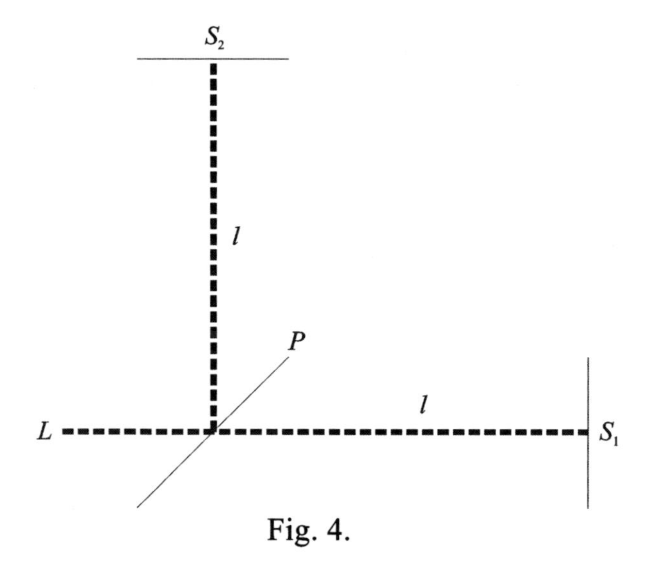

Fig. 4.

same distance l from P. If, now, the earth with the whole apparatus moves in the direction PS_1 with the velocity v, then the time which the light needs in order to go from P to S_1 and back is:

$$\frac{l}{c-v}+\frac{l}{c+v}=\frac{2l}{c}\left(1+\frac{v^2}{c^2}+\cdots\right).$$

On the other hand, the time which the light needs in order to pass from P to S_2 and back to P is:

$$\frac{l}{\sqrt{c^2-v^2}}+\frac{l}{\sqrt{c^2-v^2}}=\frac{2l}{c}\left(1+\frac{1}{2}\frac{v^2}{c^2}+\cdots\right).$$

If, now, the whole apparatus be turned through a right angle, a noticeable displacement of the interference bands should result, since the time for the passage over the path PS_2 is now longer. No trace was observed of the marked effect to be expected.

Now, how will it be possible to bring into line this result, established by repeated tests with all the facilities of modern experimental art? E. Cohn has attempted to find the necessary compensation in a certain influence of the air in which the rays are propagated. But for anyone who bears in mind the great results of the atomic theory of dispersion and who does not renounce the simple explanation which this theory gives for the dependence of the refractive index upon the color, without introducing something else in its place, the idea that a moving absolutely transparent medium, whose refractive index is absolutely = 1, shall yet have a notable influence upon the velocity of propagation of light, as the theory of Cohn demands, is not possible of assumption. For this theory distinguishes essentially a transparent medium, whose refractive index is = 1, from a perfect vacuum. For the former the velocity of propagation of light in the direction of the velocity v of the medium with relation to an observer at rest is

$$q = c + \frac{v^2}{c},$$

for a vacuum, on the other hand, $q = c$. In the former medium, Cohn's theory of the Michelson experiment predicts no effect, but, on the other hand, the Michelson experiment should give a positive effect in a vacuum.

In opposition to E. Cohn, H. A. Lorentz and FitzGerald ascribe the necessary compensation to a contraction of the whole optical apparatus in the direction of the earth's motion of the order of magnitude v^2/c^2. This assumption allows better of the introduction again of the principle of relativity, but it can first completely satisfy this principle when it appears, not as a necessary hypothesis made to fit the present special case, but as a consequence of a much more general postulate. We have to thank A. Einstein for the framing of this postulate and H. Minkowski for its further mathematical development.

Above all, the general principle of relativity demands the renunciation of the assumption which led H. A. Lorentz to the framing of his theory of a quiescent ether; the assumption of a substantial carrier of electromagnetic waves. For, when such a carrier is present, one must assume a definite velocity of a ponderable body as definable with respect to it, and this is exactly that which is excluded by the relativity principle. Thus the ether drops out of the theory and with it the possibility of mechanical explanation of electrodynamic processes, i. e., of referring them to motions. The latter difficulty, however, does not signify here so much, since it was already known before, that no mechanical theory, founded upon the continuous motions of the ether permits of being completely carried through (cf. p. 111). In place of the so-called free ether there is now substituted the absolute vacuum, in which electromagnetic energy is independently propagated, like ponderable atoms. I believe it follows as a consequence that no physical properties can be consistently ascribed to the absolute vacuum. The dielectric constant and the magnetic permeability of a vacuum have no absolute meaning, only relative. If an electrodynamic process were to occur in a ponderable medium as in a vacuum, then it would have absolutely no sense to distinguish between field strength and induction. In fact, one can ascribe to the vacuum any arbitrary value of the dielectric constant, as is indicated by the various systems of units. But how is it now with regard to the velocity of propagation of light? This also is not to be regarded as a property of the vacuum, but as a property of electromagnetic energy which is present in the vacuum. Where there is no energy there can exist no velocity of propagation.

With the complete elimination of the ether, the opportunity is now present for the framing of the principle of relativity. Obviously, we must, as a . simple consideration shows, introduce something radically new. In order that the moving observer B mentioned above (Fig. 3, p. 116) shall not see the light signal given by him traveling more slowly in his own direction of motion (with the velocity $c - v$) than in the opposite direction; (with the velocity $c + v$), it is necessary that he shall not identify the instant of time at which the light has covered the distance $c - v$ in the direction of his own motion with the instant of time at which the light has covered the distance $c + v$ in the opposite direction, but that he regard the latter instant of time as later. In other words: the observer B measures time differently from the observer A. This is a priori quite permissible; for the relativity principle only demands that neither of the two observers, shall come into contradiction with himself. However, the possibility is left open that the specifications of time of both observers may be mutually contradictory.

It need scarcely be emphasized that this new conception of the idea of time makes the most serious demands upon the capacity of abstraction and the projective power of the physicist. It surpasses in boldness everything previously suggested in speculative natural phenomena and even in the philosophical theories of knowledge: non-euclidean geometry is child's play in comparison. And, moreover, the principle of relativity, unlike noneuclidean geometry, which only comes seriously into consideration in pure mathematics, undoubtedly possesses a real physical significance. The revolution introduced by this principle into the physical conceptions of the world is only to be compared in extent and depth with that brought about by the introduction of the Copernican system of the universe.

Since it is difficult, on account of our habitual notions concerning the idea of absolute time, to protect ourselves, without special carefully considered rules, against logical mistakes in the necessary processes of thought, we shall adopt the mathematical method of treatment. Let us consider then an electrodynamic process in a pure vacuum; first, from the standpoint of an observer A; secondly, from the standpoint of an observer B, who moves relatively to observer A with a velocity v in the direction of the x-axis. Then, if A employ the system of reference x, y, z, t, and B the system of reference x', y', z', t', our first problem is to find the relations among the primed and the unprimed quantities. Above all, it is to be noticed that since both systems of reference, the primed and the unprimed, are to be like directed, the equations of transformation between corresponding quantities in the two systems must be so established that it is possible through a transformation of exactly the same kind to pass from the first system to the second, and conversely, from the second back to the first system. It follows immediately from this that the velocity of light c' in a vacuum for the observer B is exactly the same as for the observer A. Thus, if c' and c are different, $c' > c$, say, it would follow that: if one passes from one observer A to another observer B who moves with respect to A with uniform velocity, then he would find the velocity of propagation of light for B greater than for A. This conclusion must likewaise hold quite in general independently of the direction in which B moves with respect to A, because all directions in space are equivalent for the observer A. On the same grounds, in passing from B to A, c must be greater than c' for all directions in space for the observer B are now equivalent.

Since the two inequalities contradict, therefore c' must be equal to c. Of course this important result may be generalized immediately, so that the totality of the quantities independent of the motion, such as the velocity of light in a vacuum, the constant of gravitation between two bodies at rest, every isolated electric charge, and the entropy of any physical system possess the same values for both observers. On the other hand, this law does not hold for quar.tities such as energy, volume, temperature, etc. For these quantities depend also upon the velocity, and a body which is at rest for *A* is for *B* a moving body.

We inquire now with regard to the form of the equations of transformation between the unprimed and the primed coordinates. For this purpose let us consider, returning to the previous example, the propagation, as it appears to the two observers *A* and *B*, of an instantaneous signal creating an infinitely thin light wave which, at the instant at which the observers meet, begins to spread out from the common origin of coordinates. For the observer *A* the wave travels out as a spherical wave:

$$x^2 + y^2 + z^2 - c^3 t^3 = 0. \quad (66)$$

For the second observer *B* the same wave also travels as a spherical wave with the same velocity:

$$x'^2 + y'^2 + z^2 - c^2 t'^2 = 0; \qquad (67)$$

for the first observer has no advantage over the second observer. *B* can exactly with the same right as *A* assert that he is at rest at the center of the spherical wave, and for *B*, after unit time, the wave appears as in Fig. 5, while its appearance for the observer *A* after unit time, is represented by Fig. 3 (p. 116).[1]

1 The circumstance that the signal is a finite one, however small the time

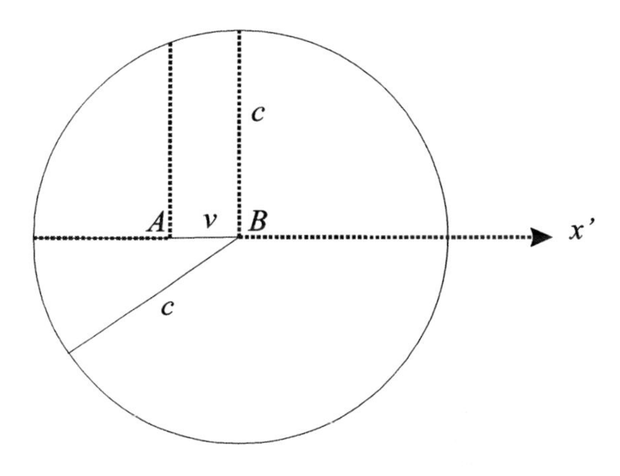

Fig. 5

The equations of transformation must therefore fulfill the condition that the two last equations, which represent the same physical process, are compatible with each other; and furthermore: the passage from the unprimed to the primed quantities must in no wise be distinguished from the reverse passage from the primed to the unprimed quantities. In order to satisfy these conditions, we generalize the equations of transformation (65), set up at the beginning of this lecture for the old mechanical principle of relativity, in the following manner:

$$x' = \kappa(x - vt), \quad y' = \lambda y, \quad z' = \mu z, \quad t' = \nu t + \rho x.$$

Here v denotes, as formerly, the velocity of the observer B relative to A and the constants $\kappa, \lambda, \mu, \nu, \rho$ are yet to be determined. We must have:

---

may be, has significance only as regards the thickness of the spherical layer and not for the conclusions here under consideration.

$$x = \kappa'(x'-v't'), \quad y = \lambda' y', \quad z = \mu' z', \quad t = v' t' + \rho' x'.$$

It is now easy to see that λ and λ' must both = 1. For, if, e. g., λ be greater than 1, then λ' must also be greater than 1; for the two transformations are equivalent with regard to the y axis. In particular, it is impossible that λ and λ' depend upon the direction of motion of the other observer. But now, since, in accordance with what precedes, $\lambda = 1/\lambda'$, each of the two inequalities contradict and therefore $\lambda = \lambda' = 1$; likewise, $\mu = \mu' = 1$. The condition for identity of the two spherical waves then demands that the expression (66):

$$x^2 + y^2 + x^2 - c^2 t^2$$

become, through the transformation of coordinates, identical with the expression (67) :

$$x'^2 + y'^2 + x'^2, - c^2 t'^2,$$

and from this the equations of transformation follow without ambiguity:

$$x' = \kappa(x - vt), \quad y' = y, \; z' = z, \quad t = \kappa\left(t - \frac{v}{c^2} x\right), \; (68)$$

wherein

$$\kappa = \frac{c}{\sqrt{c^2 - v^2}} .$$

Conversely:

$$x = \kappa(x' + vt'), y = y', z = z', t = \kappa(t' + \frac{v}{c_2} x'). \quad (69)$$

These equations permit quite in general of the passage from the system cf reference of one observer to that of the other (H. A. Lorentz), and the principle of relativity asserts that all processes in nature occur in accordance with the same laws and with the same constants for both observers (A. Einstein). Mathematically considered, the equations of transformation correspond to a rotation in the four dimensional system of reference (x, y, z, ict) through the imaginary angle arctg $(i(v/c))$ (H. Minkowski). Accordingly, the principle of relativity simply teaches that there is in the four dimensional system of space and time no special characteristic direction, and any doubts concerning the general validity of the principle are of exactly the same kind as those concerning the existence of the antipodians upon the other side of the earth.

We will first make some applications of the principle of relativity to processes which we have already treated above. That the result of the Michelson experiment is in agreement with the principle of relativity, is immediately evident; for, in accordance with the relativity principle, the influence of a uniform motion of the earth upon processes on the earth can under no conditions be detected.

We consider now the Fizeau experiment with the flowing liquid (see p. 114). If the velocity of propagation of light in the liquid at rest be again q_0, then, in accordance with the relativity principle, q_0 is also the velocity of the propagation of light in the flowing liquid for an observer who moves with the liquid, in case we disregard the dispersion of the liquid; for the color of the light is different for the moving observer. If we call this observer B and the velocity of the liquid as above, v, we may employ immediately the above formulae in the calculation of the velocity of propagation of light in the flowing liquid, judged by an observer A at the screen S. We have only to put

$$\frac{dx'}{dt'} = \dot{x}' = q_0 ,$$

and to seek the corresponding value of

$$\frac{dx}{dt} = \dot{x} .$$

For this obviously gives the velocity sought.

Now it follows directly from the equations of transformation (69) that:

$$\frac{dx}{dt} = \dot{x} = \frac{\dot{x}' + v}{1 + \frac{v\dot{x}'}{c^2}} ,$$

and, therefore, through appropriate substitution, the velocity sought in the upper tube, after neglecting higher powers in v/c and v/q_0, is:

$$\dot{x} = \frac{q_0 + v}{1 + \frac{vq_0}{c^2}} = q_0 + v(1 - \frac{q_0^2}{c^2}) ,$$

and the corresponding velocity in the lower tube is:

$$q_0 = v(1 - \frac{q_0^2}{c^2}).$$

The difference of the two velocities is

$$2v(1 - \frac{q_0^2}{c^2}) = 2v(1 - \frac{1}{n^2}),$$

which is the Fresnel coefficient, in agreement with the measurements of Fizeau.

The significance of the principle of relativity extends, not only to optical and other electrodynamic phenomena, but also to all processes of ordinary mechanics; but the familiar expression $(\frac{1}{2}mq^2)$ for the kinetic energy of a mass point moving with the velocity q is incompatible with this principle.

But, on the other hand, since all mechanics as well as the rest of physics is governed by the principle of least action, the significance of the relativity principle extends at bottom only to the particular form which it prescribes for the kinetic potential H, and this form, though I will not stop to prove it, is characterized by the simple law that the expression

$$H \cdot dt$$

for every space element of a physical system is an invariant

$$= H' \quad dt'$$

with respect to the passage from one observer A to the

other observer B or, what is the same thing, the expression $= H / \sqrt{c^2 - q^2}$ is in this passage an invariant $= H' / \sqrt{c^2 - q'^2}$

Let us now make some applications of this very general law, first to the dynamics of a single mass point in a vacuum, whose state is determined by its velocity q. Let us call the kinetic potential of the mass point for $q = 0$, H_o, and consider now the point at an instant when its velocity is q. For an observer B who moves with the velocity q with respect to the observer A, $q = 0$ at this instant, and therefore $H' = H_o$. But now since in general:

$$\frac{H}{\sqrt{c^2 - q^2}} = \frac{H'}{\sqrt{c^2 - q'^2}},$$

we have after substitution:

$$H = \sqrt{1 - \frac{q^2}{c^2}} \cdot H_0 = \sqrt{1 - \frac{\dot{x}^2 + \dot{y}^2 + \dot{z}^2}{c^2}} \cdot H_0 .$$

With this value of H, the Lagrangian equations of motion (59) of the previous lecture are applicable.

In accordance with (60), the kinetic energy of the mass point amounts to:
and the momentum to:

$$E = \dot{x}\frac{\partial H}{\partial \dot{x}} + \dot{y}\frac{\partial H}{\partial \dot{y}} + \dot{z}\frac{\partial H}{\partial \dot{z}} - H = q\frac{\partial H}{\partial q} - H = -\frac{H_0}{\sqrt{1 - \frac{q^2}{c^2}}},$$

G/q is called the transverse mass m_t, and dG/dq the longitudinal mass m_l, of the point; accordingly:

$$m_t = -\frac{H_0}{c\sqrt{c^2 - q^2}}, \quad m_l = -\frac{cH_0}{\left(c^2 = q^2\right)^{\frac{3}{2}}}.$$

For $q = 0$, we have

$$m_t = m_l = m_0 = -\frac{H_0}{c^2}.$$

It is apparent, if one replaces in the above expressions the constant H_0 by the constant m_0, that the momentum is:

$$G = \frac{m_0 q}{\sqrt{1 - \frac{q^2}{c^2}}}$$

and the transverse mass:

$$m_t = \frac{m_0}{\sqrt{1 - \frac{q^2}{c^2}}},$$

and the longitudinal mass:

$$m_l = \frac{m_0}{\left(1 - \frac{q^2}{c^2}\right)^{\frac{3}{2}}},$$

and, finally, that the kinetic energy is:

$$E = \frac{m_0 c^2}{\sqrt{1 - \frac{q^2}{c^2}}} = m_0 c^2 + \tfrac{1}{2} m_0 q^2 + \cdots,$$

The familiar value of ordinary mechanics $\frac{1}{2} m_0 q^2$ appears here therefore only as an approximate value. These equations have been experimentally tested and confirmed through the measurements of A. H. Bucherer and of E. Hupka upon the magnetic deflection of electrons.

A further example of the invariance of $H \cdot dt$ will be taken from electrodynamics. Let us consider in any given medium any electromagnetic field. For any volume element V of the medium, the law holds that $V \cdot dt$ is invariant in the passage from the one to the other observer. It follows from this that H/V is invariant; For $q = 0$, we have

$$m_t = m_l = m_0 = -\frac{H_0}{c^2}.$$

It is apparent, if one replaces in the above expressions the constant H_0 by the constant m_0, that the momentum is:

$$G = \frac{m_0 q}{\sqrt{1 - \frac{q^2}{c^2}}}$$

and the transverse mass:

$$m_t = \frac{m_0}{\sqrt{1 - \frac{q^2}{c^2}}},$$

and the longitudinal mass:

$$m_l = \frac{m_0}{\left(1 - \frac{q^2}{c^2}\right)^{\frac{3}{2}}},$$

and, finally, that the kinetic energy is:

$$E = \frac{m_0 c^2}{\sqrt{1 - \frac{q^2}{c^2}}} = m_0 c^2 + \tfrac{1}{2} m_0 q^2 + \cdots,$$

The familiar value of ordinary mechanics $\frac{1}{2} m_0 q^2$ appears here therefore only as an approximate value. These equations have been experimentally tested and confirmed through the measurements of A. H. Bucherer and of E. Hupka upon the magnetic deflection of electrons.

A further example of the invariance of $H \cdot dt$ will be taken from electrodynamics. Let us consider in any given medium any electromagnetic field. For any volume element V of the medium, the law holds that $V \cdot dt$ is invariant in the passage from the one to the other observer. It follows from this that H/V is invariant; i.e., the kinetic potential of a unit volume or the "*space density of kinetic potential*" is invariant.

Hence the following relation exists;

$$\mathfrak{E}\mathfrak{D} - \mathfrak{H}\mathfrak{B} = \mathfrak{E}'\mathfrak{D}' - \mathfrak{H}'\mathfrak{B}',$$

wherein E and H denote the field strengths and D and B

the corresponding inductions. Obviously a corresponding law for the space energy density ED + HB will not hold.

A third example is selected from thermodynamics. If we take the velocity q of a moving body, the volume V and the temperature T as independent variables, then, as I have shown in the previous lecture (p. 105), we shall have for the pressure p and the entropy S the following relations:

$$\frac{\partial H}{\partial V} = p \quad \text{and} \quad \frac{\partial H}{\partial T} = S\,,$$

Now since $V/\sqrt{c^2 - q^2}$ is invariant, and S likewise invariant (see p. 121), it follows from the invariance of $H/\sqrt{c^2 - q^2}$ that p is invariant and also that $T/\sqrt{c^2 - q^2}$ is invariant, and hence that:

$$p = p' \quad \text{and} \quad \frac{T}{\sqrt{c^2 - q^2}} = \frac{T'}{\sqrt{c^2 - q^2}}\,,$$

The two observers A and B would estimate the pressure of a body as the same, but the temperature of the body as different.

A special case of this example is supplied when the body considered furnishes a black body radiation. The black body radiation is the only physical system whose dynamics (for quasi-stationary processes) is known with absolute accuracy. That the black body radiation possesses inertia was first pointed out by F. Hasenöhrl. For black body radiation at rest the energy $E_0 = aT^4V$ is given by the Stefan-Boltzmann law, and the entropy

$S_0 = \int (dE_0 / T) = \frac{4}{3} a T^3 V$ and the pressure $p_0 = (a/3)T^4$, and, therefore, in accordance with the above relations, the kinetic potential is:

$$H_0 = \frac{a}{3} T^4 V .$$

Let us imagine now a black body radiation moving with the velocity q with respect to the observer A and introduce an observer B who is at rest ($q = 0$) with reference to the black body radiation; then:

$$\frac{H}{\sqrt{c^2 - q^2}} = \frac{H'}{\sqrt{c^2 - q'^2}} = \frac{H_0'}{c} ,$$

wherein

$$H_0' = \frac{a}{3} T'^4 V' .$$

Taking account of the above general relations between T' and T, V' and V, this gives for the moving black body radiation the kinetic potential:

$$H = \frac{a}{3} \frac{T^4 V}{\left(1 - \frac{q^2}{c^2}\right)^2} ,$$

from which all the remaining thermodynamic quantities: the pressure p, the energy E, the momentum G, the longitudinal and transverse masses m_l and m_t of the moving black body radiation are uniquely determined.

Colleagues, ladies and gentlemen, I have arrived at the conclusion of my lectures. I have endeavored to bring before you in bold outline those characteristic advances in the present system of physics which in my opinion are the most important. Another in my place would perhaps have made another and better choice and, at another time, it is quite likely that I myself should have done so. The principle of relativity holds, not only for processes in physics, but also for the physicist himself, in that a fixed system of physics exists in reality only for a given physicist and for a given time. But, as in the theory of relativity, there exist invariants in the system of physics: ideas and laws which retain their meaning for all investigators and for all times, and to discover these invariants is always the real endeavor of physical research. We shall work further in this direction in order to leave behind for our successors where possible ---lasting results. For if, while engaged in body and mind in patient and often modest individual endeavor, one thought strengthens and supports us, it is this, that we in physics work, not for the day only and for immediate results, but, so to speak, for eternity.

I thank you heartily for the encouragement which you have given me. I thank you no less for the patience with which you have followed my lectures to the end, and I trust that it may be possible for many among you to furnish in the direction indicated much valuable service to our beloved science.

Planck's Columbia Lectures

Summary of the Eighth Lecture

Planck has presented a very clear and concise description of the General Theory of Relativity. Although Einstein's theory will not be examined in great detail here, there are some interesting points that will be discussed regarding the history of the development of this theory and of the key assumptions upon which they is based.

An early hypothesis that preceded the theory of relativity was presented by physicist G. A. Fitzgerald, who argued that there is an apparent contraction in size of objects occurs that varies as a function of v^2/c^2 due to the presence of the "quiescent ether" in the universe. The assumption of an ether that is similar to a material medium was shown to be contradictory by the Fizeau experiment cited by Planck. H. A. Lorentz formulated a set of equations, based on the Fitzgerald contraction, as a function of time and space. This mathematical model allowed a resolution of the contradiction between theory and measurement, but the meaning of the equations and their parameters was up to question.

Although the Lorentz equation was based on the presence of an ether, the general relativity theory of A. Einstein, although based on the Lorentz equation, does not depend upon the presence of an ether! Einstein's conceptual description was framed into a mathematical form by H Minkowski. Planck discussed the differences between the two theories. The Fitzgerald/Lorentz theory assumes that a *quiescent ether* is the *carrier* of electromagnetic waves. The general assumption had been that the velocity of a ponderable body can be measured with respect to the ether. With the relativity principle, the dependence upon the presence of an ether drops out, along with the mechanical explanation of electrodynamic processes and the assumption that space contracts with velocity.

Another problem with the ether theory is that electrodynamic processes are continuous, which does not seem to fit the concept of quantum changes of electromagnetic energy (photons). Planck pointed out, in this lecture, that if electromagnetic process were to occur in a ponderable medium (ether), then *it would not then be possible to distinguish between field strength and induction*[10]. According to the relativity theory, which assumes a complete vacuum in which electromagnetic energy is propagated like ponderable atoms, no physical properties can be assigned to the vacuum. The velocity of propagation is not a property of the vacuum, and "where there is no energy there can be no velocity of propagation". The velocity of propagation is only dependent upon electromagnetic energy and not the medium.

In the Einstein/Minkowski model, time is allowed to vary with respect to two observers, one of which is moving with respect to the other. The Lorentz equations were modified to produce an apparent variation in the observed dimensions of spatial objects. This last assumption is the most radical. As Planck emphasized, "this new conception of time makes the most serious demands upon the capacity of abstraction and the projective power of the physicist". While Planck provided several examples of physical phenomena that are in accordance with this concept.

Another accepted assumption applies to electromagnetic radiation, which is believed to travel out as spherical waves (this assumption is also basic to the application of some of Maxwell's field equations). The

10 This issue will be discussed in detail in Chapter 10.

spherical wave equation is

$$x^2 + y^2 + z^2 - c^2t^2 = 0. \quad (66)$$

For the second observer of the wave, *x*, *y*, *z* and *t* vary, resulting in the modified wave equation,

$$x'^2 + y'^2 + z'^2 - c^2t'^2 = 0. \quad (67)$$

Only the *relative* position of one to the other affects the observation. The Lorentz equations of transformation describe the change in dimensions of the variables in the wave equation. For the first observer,

$$x' = \frac{c(x - vt)}{\sqrt{c^2 - v^2}}, \quad t' = \frac{c(t - \frac{vx}{c^2})}{\sqrt{c^2 - v^2}}, \quad (68)$$

while for the other observer,

$$x = \frac{c(x' - vt')}{\sqrt{c^2 - v^2}}, \quad t = \frac{c(t' - \frac{vx'}{c^2})}{\sqrt{c^2 - v^2}}. \quad (69)$$

In the example that was illustrated in Figure 5, only relative motion in the *x*-direction affects its apparent dimension, and the other dimensions remain unchanged as is shown in these equations.

The equations of transformation, as developed by Minkowski for a four-dimensional system (*x*, *y*, *z*, *ict*), correspond to a rotation through the imaginary angle *iv*/*c*. This mathematical model describes the novel concept of the fourth dimension of time, which has received much publicity in the general public. Uncharacteristically, Planck does not offer any physical example in which the imaginary angle appears. The question of the physical concept of rotation with respect to this imaginary angle will be discussed in the final chapter, along with an example.

Planck applied the relativity theory to the Fizeau experiment in which the difference between the velocities of the waves in the upper and lower tubes results in a change in velocity in lower tube with respect to that of the to the top tube. The resulting equation (the Fresnel coefficient) is

$$2v\left(1-\frac{q_0^{\,2}}{c^2}\right)=2v\left(1-\frac{1}{n^2}\right),$$

in which n is the index of refraction that is a measure of the speed of light in a medium. This result is in agreement with the measurements of Fizeau, thus providing a resolution of the contradiction mentioned above.

While the relativity principle extends to optical, electrodynamic and ordinary mechanics phenomena, Planck observed that the kinetic energy of a mass point

$$\frac{mq^2}{2}$$

is “incompatible with this princiṗle”. The meaning of this statement is defined by the Principle of Least Action.

According to Planck, the Principle of Least Action applies to all physical phenomena. The kinetic potential H of a space element in a physical system varies according to the law

$$H \cdot dt = H' \cdot dt'$$

with respect to the two observers in the above physical example. Therefore the following relationship also applies,

$$\frac{H}{\sqrt{c^2 - q^2}} = \frac{H'}{\sqrt{c^2 - q'^2}},$$

and

$$H = \sqrt{1 - \frac{q^2}{c^2}} \cdot H_0 .$$

The Lagrangian equations of motion (59) apply, and in accordance with equation (60) the kinetic energy of the mass is found to be

$$E = -\frac{H_0}{\sqrt{1 - \frac{q^2}{c^2}}}$$

and the momentum

$$G = -\frac{qH_0}{c\sqrt{c^2 - q^2}} .$$

The "transverse mass" is defined as G/q, and the "longitudinal mass" as dG/dq. For a system at rest (fixed positions of one observer to the other), $q = 0$, and for $H_0 = m_0$

$$m_t = m_l = m_0 = -\frac{H_0}{c^2} = -\frac{m_0}{c_0}.$$

Substituting into the above equations, the transverse mass is

$$m_t = \frac{m_0}{\sqrt{1 - \frac{q^2}{c^2}}},$$

and the longitudinal mass is

$$m_l = \frac{m_0}{\left(1 - \frac{q^2}{c^2}\right)^{\frac{3}{2}}}.$$

The two expressions are equal for $q = 0$ and both increase with velocity, but the longitudinal mass becomes predominate as q approaches the speed of light.

The kinetic energy is

$$E = \frac{m_0 c^2}{\sqrt{1 - \frac{q^2}{c^2}}},$$

which illustrates the variation from the value of ordinary mechanics,

$$E_0 = \frac{m_0 c^2}{2},$$

previously described. Therefore, in relativity theory both the energy and the mass increase with the velocity of the moving body, and the mass varies as a function of the observed direction with respect to the velocity (this directional characteristic is, I believe, more suited to field theory, rather than mechanical definitions).

The final example is the case where the body is in the form of black body radiation. Planck claimed that it is the only quasi-stationary process whose dynamics are known with absolute accuracy. For black body radiation at rest, the energy is described by the Stefan-Boltzmann law

$$E_0 = aT^4V,$$

the entropy is

$$S_0 = \int\left(\frac{dE_0}{T}\right) = \frac{4}{3}aT^2V,$$

and the pressure is

$$p_0 = \frac{aT^4}{3}.$$

Therefore, the kinetic potential of a stationary body is

$$H_0 = \frac{aT^4V}{3}.$$

and for a moving black body,

$$H = \frac{aT^4V}{3\left(1-\frac{q^2}{c^2}\right)^2}.$$

The thermodynamic processes, pressure, energy, momentum, and longitudinal and transverse masses, of the moving black body radiation are all determined from this equation. Therefore, the kinetic potential is called an "invariant".

In his closing statements, Planck clearly regarded the invariants in the system of physics as being among the most important advances in the state of physics, to which his *Blackbody Radiation Theory* and the *Principle of Least Action* most certainly belong.

Chapter 9

Abridged Lectures

Planck's eight lectures are presented in this chapter. The translated text of the original document evidently was done in a literal manner, preserving some of the German style of sentence parsing, which is sometimes difficult to follow. Here, the text has been written in a modern English style for easier reading. In addition, some of the original mathematics has been written in a manner that is intended to be easier to understand. These lectures have been abridged, but not extensively. For instance, Chapter 2, includes seven chemical applications that are written in detail with very little modification, which was done because these applications represent the various correlations between energy state models and real world measurements. Planck's successful results lend greater credence to the belief that his theory correlates with the physical world.

Planck's Columbia Lectures

FIRST LECTURE: REVERSILITY AND IRREVERSIBILITY

Planck began his presentation by establishing the principles upon which his theory is based. He exhorted the idea of the unification of our system of physics, in which the main effort would be toward the elimination of the individuality of the particular physicist and the establishment of a common system for all physicists. Sense perceptions in physical research form the source of all our knowledge and lead to the point of departure from original conceptions. All physical ideas depend upon measurements, and mathematics is the chief tool with which the material is worked. However, he said that depending too much upon theory is dangerous, and the strongest impetus in the development of science is by means of the forces employed in the struggle against old points of view.

The unification of the system of physics would result in merging two categories: *mechanics* and *electrodynamics*. The domain of mechanics includes acoustics, phenomena in material bodies and chemical phenomena. The domain of electrodynamics includes the physics of the ether, magnetism, optics and radiant heat. However, he said that one domain cannot be sharply differentiated from the other, asking: "Is light a mechanical or electrodynamic process?" and "To which process shall be assigned the laws of motion of electrons?". The individual electron can be said to belong to the electromagnetic domain, and yet the analysis of Lorentz indicates that electrons belong to the kinetic theory of gases[2]. The extent of these two subjects overlapped as far back as the early 18th century, and the fundamental problem

2 The electron is both electrodynamical and mechanical? I agree.

was the amalgamation of the physics of *ether* and *matter*.

The first step was the employment of the Principle of Conservation of Energy and the First Law of Thermodynamics, which states that energy can only be transformed and not created. In a frictional process, energy is transformed into heat, and the remaining energy of the system is always less than the energy supplied. This law was proved by experiment, and this principle is leads to the impossibility of perpetual motion of the first kind.

The Second Law of Thermodynamics imposes a further limitation that allows only certain transformations subject to certain conditions. This law is related to the impossibility of perpetual motion of the second kind. Heat can be transformed into energy, as in the case of a gas that expands while doing work. When the gas cools, heat loss is lost in the process. For a periodically operating motor, in which the pressure of a heated expanding gas raises a load, the gas is cooled and must again be heated, which requires additional energy. In Planck's hypothetical example, a motor can be constructed that raises a load and at the same time acts as a refrigerating machine. If the ocean could be utilized as the heat reservoir to restore the heat loss, it would involve no cost to run it. This example fits W. Ostwald's definition of perpetual motion of the second kind. But even if such a machine could be constructed, it would not fit Planck's definition of the second law, for within the gas, there are individual variations in pressure for its macroscopic elements that cannot be completely restored in the process.

The impossibility of perpetual motion of the first kind leads to the principle of the conservation of energy, while the impossibility of perpetual motion of the second kind leads to the second law of thermodynamics. At this point, he introduced the concepts of *reversibility* and *irreversibility*. Irreversible systems lose heat in the process and cannot be exactly restored to their prior condition. For example, a body that is at a fixed temperature may constantly lose energy by heat radiation, but the states of all of the molecules within each macroscopic region experience considerable variations and cannot all be restored to their prior conditions. The states of an individual molecule, however, can be restored under suitable conditions, and it is therefore a reversible system. Planck's general law is: *there are processes in nature which in no possible way can be made completely reversible.* A process that cannot be made completely reversible is an irreversible process, and the second law of thermodynamics applies to irreversible processes. Examples of reversible processes are the motions of the planets, frictional flow and undamped vibrations.

Heat conduction is an irreversible process. A body exhibiting heat loss passes from one state to another. Planck claimed that nature has a "preference" for one state over another. Reversible processes are the limiting case in which the preferences are equal. The measure of this preference of nature is by a quantity called "entropy". Clausius had introduced the concept of entropy, which is defined in terms of the quantity of heat exchanged in going from one temperature state to another. The entropy, S, is decreased by heat loss in the amount Q/T, and $\Delta S = Q/T$. For an isothermal cyclical process, heat is produced and work is consumed. In the conduction of heat from a warmer body to a colder body, the entropy of the warmer body decreases, while that of the colder body increases. The entropy of the warmer body will always be greater or equal to that of the colder body, and therefore the sum of the entropies of the two bodies will always be greater than zero.

In the limiting case of a reversible isothermal process, the work consumed is zero, and there is no heat produced. In this case, the sum of the entropies is zero. These results are in accordance with the Second Law of Thermodynamics. The distinctions between reversible and irreversible systems are utilized in classifying all physical processes.

Planck's Columbia Lectures

SECOND LECTURE: Thermodynamic States of Equilibrium in Solutions.

There Is a great deal of information presented in this lecture. Planck continues building the foundation of his theory as based on established laws and physical measurements. In each example, he accounts for every molecule of every weight, number and type, all of the pressures and temperatures to which they are subjected, their energy, mass and volume, their gas constants, the energy states that they attain, and their aggregations.

For these applications, Planck chose to utilize certain chemical reactions upon which extensive measurements have been made. Characterizations are made possible by a few restrictions, the use of dilute solutions in these applications being most useful for this purpose. The relationship of the gas constants of liquids and solids and gases to the states of equilibrium is interesting. These enumerated constants are the "parameters" of mathematical equations, which are often generally considered the least interesting terms in a purely mathematical analysis. Clearly, the parameters of equations that represent physical systems can be quite meaninful.

The derivation of the most important seven of the eleven numbered equations that he used in his analysis will be emphasized in this summary. Not all of the mathematical formulas are necessary for comprehension of the methodology. In fact, just two primary equations were utilized in all seven applications.

The state of equilibrium is fundamental to the definition of the molecular states, as are their entropies. A fundamental law of thermodynamics is that an isolated system, whose entropy S reaches a maximum value, is in a state of equilibrium, and he gives credit for this law to the physicist John Willard Gibbs. When a system at a given pressure p and temperature T is completely isolated from external excitation, then the entropy can only increase,

$$dS > 0,$$

which continues until a condition of equilibrium is reached. This relationship applies to a system of any number of constituents in an arbitrary number of phases, which is a very general characterization.

Applying the first law of thermodynamics to an isothermal-isobaric system ($T = \text{const.}, p = \text{const.}$),

$$dS + dS_0 > 0,$$

in which the entropy of the bodies surrounding the system is denoted by S_0. The change in entropy of the surroundings decreases in the process and is a function of the heat loss Q,

$$dS_0 = -\frac{Q}{T}.$$

Therefore, the change in entropy is proportional to the heat loss and inversely proportional to the absolute temperature.

In accordance with the first law of thermodynamics, the heat loss is function of the energy U, pressure and volume of the system,

$$Q = dU + pdV.$$

In this equation, the amount of heat loss is equal to the change in energy plus the pressure times the change in volume of the system. This differential equation is substituted into the entropy equation,

$$dS - \frac{dU + pdV}{T} > 0,$$

and since p and T are constant,

$$d\left(S - \frac{U + pV}{T}\right) > 0.$$

For every change of state of a physical system (under the conditions stated above), the entropy increases, and therefore the value of the bracketed term in the above equation increases. The energy state Φ of an isothermal-isobaric system is defined as

$$\Phi \triangleq S - \frac{U + pV}{T}, \quad (1)$$

which increases for every change in state,

$$d\Phi > 0.$$

Therefore the system undergoes a series of state changes, and Φ increases until an absolutely stable state of equilibrium is reached and no further change is possible,

$$\delta\, \Phi = 0. \quad (2)$$

Equation (2) represents the differential of Φ alone, regardless of how the other variables change, and the system reaches a stable state when Φ reaches a maximum.

The expression (1) for Φ is a function of the

independent variable for each phase of the system, and equation (2) defines the conditions for stability and equilibrium. Equation (1) is linear and homogeneous in S, U and V, and Φ is the sum of the quantities that represent the phases of the system. If the expression for Φ is known, then equation (2) can be utilized to determine the conditions for stable equilibrium. This is the case for dilute solutions, which is the reason for using this choice in the applications that are analyzed.

In thermodynamics, a "solution" is composed of a series of different molecular complexes, each of which is represented by a molecular number, and one particular molecular number is great with respect to the remaining complexes. The state is determined by the pressure p and the temperature T, as described by the above equation, and the volume V and the energy U are linear functions of the molecular numbers of the constituents:

$$V = n_0 v_0 + n_1 v_1 + n_2 v_2 \cdots = n_0 v_0 + \sum_{k=1}^{N} n_k v_k \, ,$$

$$U = n_0 u_0 + n_1 u_1 + n_2 u_2 \cdots = n_0 u_0 + \sum_{k=1}^{N} n_k u_k \, .$$

In these equations, the number of the molecules n_0 in the predominate complex is much greater than the numbers of the molecules in the other complexes, which is the case for a dilute solution. The expression to the far right of the equations is a compact mathematical expression for a series of terms. These expressions encompass a very broad range of possibilities for various dilute solutions.

The next step is to apply the entropy equation. At equilibrium, $\delta\Phi = 0$, and from the general equation of entropy (1)

$$S = \frac{U + pV}{T}.$$

The state is determined by the changes in pressure p and temperature T, and the differential entropy is

$$dS = \frac{dU + pdV}{T},$$

and upon substitution of the above series equations,

$$dS = n_0 \frac{du_0 + pdv_0}{T} T + \sum_{k=1}^{N} n_k \frac{du_k + pdv_k}{T}.$$

Each term in the above equation depends only upon p and T and are complete differentials,

$$\frac{d\,u_0 + p\,dv_0}{T} = ds_0, \quad \frac{d\,u_k + p\,dv_k}{T} = ds_k. \qquad (3)$$

Integrating the above series equation term-by-term,

$$S = \sum_{k=0}^{N} n_k s_k + C,$$

in which the individual entropies s_k depend upon p and T, and the constant C may depend only upon the molecular numbers n_k.

The entropy is a definite value when a solution reaches the state where it consists of a mixture of ideal gases after undergoing an appropriate increase of temperature pressure and decrease in pressure. Every solution has this property, and the integration constant is

$$C = -R\sum_{k=0}^{N} n_k \log c_k$$

in accordance with the work of Gibbs. In this equation, R is the absolute gas constant, and c_k denotes the molecular concentrations of the gases,

$$c_k = \frac{n_k}{\sum_k n_k}.$$

Consequently, the general entropy of a dilute solution is

$$S = \sum_{k=0}^{N} n_k \left(s_k - R\log c_k\right).$$

Substituting the result into equation (1), results in the overall phase equation (4)

$$\Phi = \sum_{k=0}^{N} n_k(\varphi_k - R\log c_k). \quad (4)$$

The gas constant, R, in this equation arises from the fact that every solution passes into a state of a *mixture of ideal gases* as the temperature is increased and the pressure decreased appropriately (per Gibbs).

Equation (5) applies to each of the k individual phases of each molecular complex

$$\varphi_k = s_k - \frac{u_k + p v_k}{T}, \qquad (5)$$

in which all quantities depend only upon p and T.

Thus equation (1), which represents the phase state Φ and equation (2), the condition for its maximum, were used to determine the expressions for a system in thermodynamic equilibrium. These equations were characterized using the laws for dilute solutions. The next step is to find the general law of equilibrium and apply to a series of special applications.

Planck defined the complex of homogeneous phases of a material system in terms of the individual phases for all of the molecules of each molecular complex,

$$n_0 m_0, n_1 m_1, \cdots, \quad n_0' m_0', n_1' m_1', \cdots, n_0'' m_0'', n_1'' m_1'', \cdots, \quad \cdots .$$

The n_k are the molecular numbers, and the m_k are the molecular weights. The terms that are primed represent the individual phases. When each phase contains only a single molecular complex, it represents an absolutely pure substance in which case each phase represents a dilute solution, and the concentrations of all of the dissolved substances is zero.

If an isobaric-isothermal change produces a simultaneous change in the molecular numbers, then, in accordance with equation (2) for equilibrium under constant temperature and pressure, the individual phases of the molecular complexes,

$$\Phi + \Phi' + \Phi'' + \cdots ,$$

is a maximum. Then, in accordance with the above terminology, after first substituting equation (5) into equation (4) results in the change in phase state

$$\sum_{k=0}^{N}(\varphi_k - R\log c_k)\delta n_k = 0,$$

which zero at equilibrium and extends over all phases of the system. The ratios of the δn's in the above equation are proportional to the ratios of the v's,

$$\delta n_0 : \delta n_1, \cdots, \quad \delta n_0' : \delta n_1', \cdots, \quad \delta n_0'' : \delta n_1'', \cdots =$$
$$v_0 : v_1, \cdots, \quad v_0' : v_1', \cdots, \quad v_0'' : v_1'', \cdots, \quad .$$

These ratios are simple integer positive numbers that vary with the molecular complex under consideration, and the condition for equilibrium is

$$\sum v_n \log c_n = \frac{1}{R}\sum v_n \varphi_n = \log K, \quad (6)$$

where c_n are the molecular concentrations, and the individual phases φ_n are functions of pressure p and temperature T.

The final step in this development was to obtain an expression for the variations in K produced by heat exchange and volumetric changes. Differentiating with respect to pressure and then temperature,

$$\frac{\partial \log K}{\partial p} = \frac{1}{R}\sum_{k=0}^{N} v_k \frac{\partial \varphi_k}{\partial p},$$
$$\frac{\partial \log K}{\partial T} = \frac{1}{R}\sum_{k=0}^{N} v_k \frac{\partial \varphi_k}{\partial T}.$$

Then forming the differentials of equation (5) for changes in p and T and substituting the differentials of equation (3),

$$\frac{du_k + pdv_k}{T} = ds_k,$$

and hence

$$\frac{\partial \varphi_k}{\partial p} = -\frac{v_k}{T}, \quad \frac{\partial \varphi_k}{\partial T} = -\frac{u_k + pv_k}{T^2}.$$

Substituting this result into the above series differential equations for the variations in K with p and T,

$$\frac{\partial \log K}{\partial p} = -\frac{1}{RT} \cdot \Delta V, \quad \frac{\partial \log K}{\partial T} = -\frac{\Delta Q}{RT^2}, \quad (7)$$

where ΔV is the total change in the volume of the system and ΔQ is the heat communicated to it from the outside by the isobaric-isothermal exchange in going from one condition of equilibrium to another.

Planck thus derived a set of equations for changes in the state of a system that is based on the principle of entropy and the laws of thermodynamics. These equations were applied to seven practical applications, of which seven equations [mostly just two of them, (6) and (7)] are used for most of the analysis in the following seven applications.

I. *Electrolytic Dissociation of Water*

In this application, water molecules are split into two ions in a single-phase transformation

$$n_0 H_2 O \rightarrow n_1 H^+ + n_2 H^- O.$$

Therefore, the molecular ratios in equation (6) are simple unit integers, since a water molecule v_0, is lost by dissociation into two ions (H^+ and H^+O),

$v_0 = -1, \qquad v_1 = +1 \qquad v_2 = +1.$

The equation on the left represents the original water molecule, which is lost by dissociation, into the two ions that are represented by the terms on the right. The resulting approximate molecular concentrations are

$c_0 = 1, \qquad c_1 = c_2 .$

Substituting into equation (6),

$$-\log c_0 + \log c_1 + \log c_2 = \log K,$$

and therefore,

$$2 \log c_1 = \log K.$$

The ion concentration c_1 is a function of temperature, as is seen by applying equation (7),

$$2 \frac{\partial \log c_1}{\partial T} = \frac{\Delta Q}{RT^2}.$$

The heat of dissociation of a molecule of H_2O into its two ions H^+ and $H^- O$ at a given temperature was determined by measurement (Wormann)

$$\Delta Q = 27{,}857 - 48.5T \text{ gm-cal.}$$

and is in accordance with the heat of ionizaion in the neutralization of a strong univalent base and acid in a dilute aqueous solution (Arrhenius).

Substituting the value of the gas constant into equation (7),

$$\frac{\partial \log c_1}{\partial T} = \frac{1}{2 \cdot 1.985}\left(\frac{27.857}{T^2} - \frac{48.5}{T}\right).$$

Integrating with respect to temperature,

$$\log_{10} c_1 = -\frac{3047.3}{T} - 12.125 \log_{10} T + \text{const.},$$

which agrees with agrees with the measurements of the electric conductivity of water at different temperatures (Kohlrausch and Heydweiller, Noyes and Lunden).

II. Dissociation of a Dissolved Electrolyte.

This application is a bit more complex, involving an aqueous solution of acetic acid that also forms ions, similar to the previous example:

$$n_0 H_2O, \quad n_1 H_4C_2O_2, \quad n_2 H^+, \quad n_3 H_3C_2^- O_2.$$

A single molecule of acetic acid $H_4C_2O_2$ separates into two ions, and the ratios are again unit numbers.

$$\nu_0 = 0, \quad \nu_1 = -1, \quad \nu_2 = 1, \quad \nu_3 = 1$$

(note that the water molecule did not dissociate). These simple numbers are substituted into equation (6), the antilog is taken, and a simple ratio is obtained. At equilibrium, equation (6) becomes:

$$- \log c_1 + \log c_2 + \log c_2 = \log K,$$

and since $c_2 = c_3$,

$$\frac{c_2^{\,2}}{c_1} = K.$$

The total number of the undissociated and dissociated acid molecules is independent of the degree of dissociation, and so the sum $(c_1 + c_2) = c$ is known . Therefore c_1 and c_2 may be calculated from K and c. Because of the connection between the degree of dissociation and electrical conductivity of the solution, it is possible to test the equation of equilibrium. Applying the electrolytic dissociation theory of Arrhenius, the ratio of the molecular conductivity λ of the solution in any dilution to the molecular conductivity λ of the solution in infinite dilution is:

$$\frac{\lambda}{\lambda_\infty} = \frac{c_2}{c_1 + c_2} = \frac{c_2}{c}$$

(electric conduction is accounted for by the dissociated molecules only). In accordance with the above two equations,

$$\frac{\lambda^2 c}{\lambda_\infty - \lambda} = K \cdot \lambda_\infty = \text{const.}$$

As c decreases without limit, λ increases to λ_∞.

This is Ostwald's "law of dilution" for binary electrolytes, which has been confirmed in numerous cases by experiment, as in the case of acetic acid. The dependence of the degree of dissociation upon the temperature is similar to the example above for the dissociation of water.

III. *Vaporization or Solidification of a Pure Liquid.*

In this application, a pure liquid either vaporizes or solidifies, resulting in two phases in which the number of molecules in each phase can differ. In equilibrium the system consists of two phases, one liquid, and one gaseous or solid:

$$n_0 m_0 \mid n_0' m_0'.$$

Each phase contains only a single molecular complex.

If a liquid molecule evaporates or solidifies, then

$$v_0 = -1, \quad v_0' = \frac{m_0}{m_0'}, \quad c_0 = 1, \quad c_0' = 1\,.$$

The condition for equilibrium, in accordance with (6), is:

$$0 = \log K. \qquad (8)$$

This equation therefore expresses the law of dependence of the pressure of vaporization (or melting pressure) upon the temperature, or vice versa. The quantity K depends only upon p and T, and differentiating with respect to these variables,

$$0 = \frac{\partial \log K}{\partial p} dp + \frac{\partial \log K}{\partial T} dT ,$$

and applying equation (7):

$$0 = -\frac{\Delta V}{T} dp + \frac{\Delta Q}{T^2} dT.$$

For the molecular volumes of the two phases, v_0 and v_0', the change in volume is

$$\Delta V = \frac{m_0 v_0'}{m_0'} - v_0 ,$$

and similarly, the change in heat is:

$$\Delta Q = T(\frac{m_0 v_0'}{m_0'} - v_0) \frac{dp}{dT} ,$$

or, referred to unit mass:

$$\frac{\Delta Q}{m_0} = T(\frac{v_0'}{m_0'} - \frac{v_0}{m_0}) \cdot \frac{dp}{dT} ,$$

which is the well-known formula of Carnot and Clapeyron.

In summary, he used both terms in equation (7) to yield an expression in terms of volume, temperature, pressure and heat and then finds the molecular volumes of the two phases. Upon re-arranging this equation he obtained the "...well-known formula of Carnot and Clapeyron" for the amount of heat change required to separate the molecules into the two phases as a function of temperature and pressure.

IV. *The Vaporization or Solidification of a Solution of Non-Volatile Substances.*

This application is the vaporization or solidification of a solution of non-volatile substances, such as aqueous salt solutions. The second phase is either gaseous or solid, containing only a single molecular complex. The system parameters are :

$$n_0 m_0,\ n_1 m_1,\ n_2 m_2, \cdots \mid n_0'\ m_0' \ .$$

Only one of the molecular complexes vaporizes or solidifies, so the ratio numbers of the other complexes are zero, and the change is represented by:

$$\nu_0 = -1, \quad \nu_1 = 0, \quad \nu_2 = 0, \cdots \nu_0' = \frac{m_0}{m_0'},$$

and applying the condition of equilibrium of equation (6),

$$-\log c_0 = \log K \ .$$

For small quantities of higher order, the molecular concentration is:

$$c_0 = \frac{n_0}{n_0 + n_1 + n_2 + \ldots} = 1 - \frac{n_1 + n_2 + \ldots}{n_0},$$

and the result is simply:

$$\frac{n_1 + n_2 + \cdots}{n_0} = K \ . \qquad (9)$$

Comparing this result with equation (8) in example III, shows that the solution of a foreign substance results in a small proportionate departure from the law of vaporization or solidification of the pure solvent. At a fixed pressure p, the boiling point or the freezing point T of the solution is different than that (T_0) for the pure solvent, or: at a fixed pressure T the vapor pressure or solidification pressure p of the solution is different from that (p_0) of the pure solvent. The departure in both cases is:

1. If T_0 is the boiling (or freezing temperature) of the pure solvent at the pressure p, then, in accordance with (8):

$$(\log K)_{T=T_0} = 0,$$

and by subtraction of (9):

$$\log K - (\log K)_{T=T_0} = \frac{n_1 + n_2 + \cdots}{n_0},$$

Since T is little different from T_0, with the application of equation (7),

$$\frac{\partial \log K}{\partial T}(T - T_0) = \frac{\Delta Q}{R T_0^2}(T - T_0) = \frac{n_1 + n_2 + \cdots}{n_0},$$

and therefore:

$$(T - T_0) = \frac{n_1 + n_2 + \cdots}{n_0} \cdot \frac{R T_0^2}{\Delta Q} \qquad (10)$$

This is the law for the raising of the boiling point or for the lowering of the freezing point, first derived by van't Hoff.

2. If p_0 is the vapor pressure of the pure solvent at the temperature T, then, in accordance with (8):

$$(\log K)_{p=p_0} = 0,$$

and by subtraction of (9):

$$\log K - (\log K)_{p=p_0} = \frac{n_1 + n_2 + \cdots}{n_0}.$$

Since p and p_0 are nearly equal, applying equation (7),

$$\frac{\partial \log K}{\partial p}(p - p_0) = -\frac{\Delta V}{RT}(p - p_0) = \frac{n_1 + n_2 + \cdots}{n_0}.$$

Therefore, if ΔV is equal to the volume of the gaseous molecule produced in the vaporization of a liquid molecule, then

$$\Delta V = \frac{m_0}{m_0'} \frac{RT}{p},$$

and

$$\frac{p_0 - p}{p} = \frac{m_0'}{m_0} \cdot \frac{n_1 + n_2 + \cdots}{n_0}.$$

This is the law of relative depression of the vapor pressure, first derived by van't Hoff. The factor m_0'/m_0 is frequently left out in this formula, but this is not allowable when m_0 and m_0' are unequal (as, e. g:, in the case of water).

V. *Vaporization of a Solution of Volatile Substances.*

(E. g., a Sufficiently Dilute Solution of Propyl Alcohol in Water.)

This application is similar to the previous application, except that he now considers the vaporization of a solution of volatile substances, which in this case consists of a dilute solution of propyl alcohol in water. Planck uses the method that he used in application IV in deriving equation (10) from equation (6). He allows for a more complex system, consisting of two phases, which is represented by:

$$n_0m_0,\ n_1m_1,\ n_2m_2,\quad |\ n_0'm_0',\ n_1'm_1',\ n_2'm_2',\quad ,$$

for which the subscript "$_0$" refers to the solvent and the other subscripts $_{1, 2, 3}$ . . . refer to the various molecular complexes of the dissolved substances. The addition of primes in the case of the molecular weights (m_0', m_1', m_2' . . .) allows that the various molecular complexes in the vapor may possess different molecular weights.

For a system that may experience various sorts o3f changes, there are also various conditions of equilibrium, each of which relates to a definite type of transformation. The first that change to be considered consists in the vaporization of the solvent. In accordance with our3333 scheme of notation:

$$\nu_0=-1, \nu_1=0, \nu_2=0, \cdots \nu_0'=\frac{m_0}{m_0'}, \nu_1'=0, \nu_2'=0, \cdots,$$

and, therefore, the condition of equilibrium (6) is:

$$-\log c_0+\frac{m_0}{m_0'}\log c_0'=\log K,$$

or, by substituting

$$c_0 = 1 - \frac{n_1 + n_2 + \cdots}{n_0} \quad \text{and} \quad c_0' = 1 - \frac{n_1' + n_2' + \cdots}{n_0'}, \qquad (33)$$

then

$$\frac{n_1 + n_2 + \cdots}{n_0} - \frac{m_0}{m_0'} \cdot \frac{n_1' + n_2' + \cdots}{n_0'} = \log K.$$

Then applying these results to equation (9), results in an equation similar to (10):

$$T - T_0 = \left(\frac{n_1 + n_2 + \cdots}{n_0} - \frac{n_1' + n_2' + \cdots}{n_0'} \right) = \log K.$$

Here ΔQ is the heat effect in the vaporization of one molecule of the solvent and, therefore, $\Delta Q/m_0$ is the heat effect in the vaporization of a unit mass of the solvent.

Here again, the solvent always occurs in the formula through the mass only, and not through the molecular number or the molecular weight. However, in the case of the dissolved substances, the molecular state is characterized by their influence upon vaporization.

Finally, the formula contains a generalization of the law of van't Hoff, stated above for the raising of the boiling point. Here the number of dissolved molecules in the liquid is replaced by the difference between the number of dissolved molecules in unit mass of the liquid and in unit mass of the vapor. According as the unit mass of liquid or the unit mass of vapor contains more dissolved molecules, the boiling point of the solution is raised or lowered. In the limiting case, when both quantities are equal, and the mixture therefore boils without changing, the change in boiling point becomes equal to zero. There are corresponding laws holding for the change in the vapor pressure.

In the case of the vaporization of a dissolved molecule,

$$\nu_0 = 0, \nu_1 = -1, \nu_2 = 0 \cdots, \nu_0' = 0, \nu_1' = \frac{m_1}{m_1'}, \nu_2' = 0, \cdots$$

and, in accord3ance with (6), for the condition of equilibrium:

$$-\log c_1 + \frac{m_1}{m_1'} \log c_1' = \log K,$$

or:

$$\frac{c_1'^{\frac{m_1}{m_1'}}}{c_1} = K.$$

This equation expresses the "Nernst law of distribution". If the dissolved substance possesses the same molecular weight ($m_1 = m_1'$) in both phases, then, in a state of equilibrium a fixed ratio of the concentrations c_1 and c_1' in the liquid and in the vapor exists, which depends only upon the pressure and temperature. However, if the dissolved substance polymerises, to some degree, in the liquid, then the relation demanded in the last equation appears in place of the simple ratio.

Employing equations (6) and (7), he was able to determine the heat lost in the precipitation of the succinic acid at room temperature from the heat required at two other temperatures. The results correlate with the measurements of Berthelot. This method is also applicable to the absorption of a gas, and he used CO_2 as and example to derive the concentration of the dissolved gas as a function of the pressure of the free gas above the solution which is in accordance with the law of Henry and Bunsen.

VI. *The Dissolved Substance only Passes over into the Second Phase.*

This case is in a certain sense a special case of the one preceding. It applies to the solubility of a slightly soluble salt in water, first investigated by van't Hoff, e. g., succinic acid. For this system,

$$n_0 H_2 0,\ n_1 H_6 C_4 O_4 \mid n_0'\ H_6 C_4 O_4,$$

in which the small dissociation of the acid solution is regarded as negligible. The concentrations of the individual molecular complexes are:

$$c_0 = \frac{n_0}{n_0 + n_1}, \quad c_1 = \frac{n_1}{n_0 + n_1}, \quad c_0' = \frac{n_0'}{n_0'} = 1.$$

For the precipitation of solid succinic acid,

$$\nu_0 = 0, \quad \nu_1 = -1, \quad \nu_0' = 1,$$

and, therefore, from the condition of equilibrium expressed by equation (6)

$$-\log c_1 = \log K,$$

and then from equation (7)

$$\Delta Q = -RT^2 \frac{\partial \log c_1}{\partial T}.$$

This equation was used by van't Hoff to calculate the heat of solution ΔQ from the solubility of succinic acid at 0° and at 8:5 °C. The corresponding numbers were 2.88 and 4.22 in an arbitrary unit. The result of this approximation is:

$$\frac{\partial \log c_1}{\partial T} = \frac{\log_e 4.22 - \log_e 2.88}{8.5} = 0.04494,$$

from which for $T = 273$:

$$\Delta Q = -1.98 \cdot 273^2 . 0.04494 = -6{,}600 \text{ cal.},$$

which is the heat given up to the surroundings in the precipitation of a molecule of succinic acid. Berthelot found, through direct measurement, 6,700 calories for the heat of solution.

The absorption of a gas also comes under this category, e. g. carbonic acid in a liquid, such as water, of relatively unnoticeable smaller vapor pressure at not too high a temperature. For this system,

$$n_0 H_2 0,\ n_1 CO_2 \mid n_0' CO_2.$$

The vaporization of a molecule CO_2 corresponds to the values

$$\nu_0 = 0, \quad \nu_1 = -1, \quad \nu_0' = 1.$$

The condition of equilibrium is therefore again:

$$-\log c_1 = \log K,$$

and at a fixed temperature and a fixed pressure the concentration c_1 of the gas in the solution is constant.

The change of the concentration with p and T is obtained through substitution in equation (7):

$$\frac{\partial \log c_1}{\partial p} = \frac{\Delta V}{RT}, \quad \frac{\partial \log c_1}{\partial T} = -\frac{\Delta Q}{RT^2}.$$

ΔV is the change in volume of the system that occurs in the isobaric-isothermal vaporization of a molecule of CO_2, and ΔQ is the quantity of heat absorbed in the process from outside. Now, since ΔV represents the approximate volume of a molecule of gaseous carbonic acid:

$$\Delta V \cong \frac{RT}{p},$$

and then

$$\frac{\partial \log c_1}{\partial p} = \frac{1}{p},$$

which upon integrating yields:

$$\log c_1 = \log p + \text{const.}, \qquad c_1 = C \cdot p.$$

Therefore, the concentration of the dissolved gas is proportional to the pressure of the free gas above the solution (law of Henry and Bunsen). The factor of proportionality C, which furnishes a measure of the solubility of the gas, depends upon the heat effect in the same manner as in the previous example.

A number of important relations are easily derived as by-products of those found above, e. g., the Nernst laws concerning the influence of solubility, the Arrhenius theory of isohydric solutions, etc. All such may be obtained through the application of the general condition of equilibrium defined by equation (6).

VII. *Osmotic Pressure.*

This application is somewhat different in that Planck derives the osmotic pressure between a dilute solution of a dissolved substance and a pure solution. The same general methods apply in this case. For this system,

$$n_0 m_0,\ n_1 m_1,\ n_2 m_2, \cdots \mid n_0' m_0 .$$

The condition of equilibrium is again expressed by equation (6) for a change of state in which the temperature and the pressure in each phase is maintained constant. However, for the two phases on either side of the membrane, the ordinary hydrostatic pressure p in the first phase may be different from the pressure p' in the second phase.

The proof that equation (6) is valid for this application is similar to the prior example, utilizing the principle of increase of entropy. In this more general case, the external work in a given change is represented by the sum $pdV + p'dV'$, where V and V' denote the volumes of the two individual phases, while before V denoted the total volume of all phases. Accordingly, we use the following equation instead of instead of equation

(7), to express the dependence of the constant K in (6) upon the pressure:

$$\frac{\partial \log K}{\partial p} = -\frac{\Delta V}{RT}, \quad \frac{\partial \log K}{\partial p'} = -\frac{\Delta V'}{RT}. \qquad (11)$$

The following symbolic representations,

$$\nu_0 = -1, \quad \nu_1 = 0, \quad \nu_2 = 0, \quad \cdots, \quad \nu_0' = 1,$$

express the fact that a molecule of the solvent passes out of the solution through the membrane into the pure solvent. Then in accordance with equation (6),

$$-\log c_0 = \log K,$$

and

$$c_0 = 1 - \frac{n_1 + n_2 + \cdots}{n_0}, \quad \frac{n_1 + n_2 + \cdots}{n_0} = \log K.$$

Here K depends only upon T, p and p'. With a pure solvent were present on both sides of the membrane, $c_0 = 1$, and $p = p'$, and therefore,

$$(\log K)_{p=p'} = 0.$$

By subtraction of the last two equations:

$$\frac{n_1 + n_2 + \cdots}{n_0} = \log K - (\log K)_{p=p'} = \frac{\partial \log K}{\partial p}(p - p')$$

and in accordance with (11):

$$\frac{n_1 + n_2 + \cdots}{n_0} = -(p - p')\cdot\frac{\Delta V}{RT}.$$

Here ΔV denotes the change in volume of the solution due to the loss of a molecule of the solvent ($\nu_0 = -1$). The approximate volume of the whole solution is then

$$-\Delta V \cdot n_0 = V,$$

and

$$\frac{n_1 + n_2 + \cdots}{n_0} = (p - p')\cdot\frac{\Delta V}{RT}.$$

This equation contains the well known law of osmotic pressure due to van't Hoff, wherein the difference $(p - p')$ is the osmotic pressure of the solution.

We have just experienced some of the relationships between the mathematical definition of state space and the measurements of the real world. The validity of Planck's state equation (6) for equilibrium is clearly well established by applying the thermodynamic equation (7) to these seven applications and comparing the results to previous laboratory measurements. The state equation is based on special reversible isothermal cycles by the condition of equilibrium, and the difficulty of the complex analysis of the corresponding differential equations that apply to oscillators is thereby avoided.

The system of equations that Planck developed can be applied to complex systems of molecular aggregations, which include various types of molecules of different numbers and masses and in multiple aggregations as was seen in most of the above applications. The advantage of dealing with energies and their quantum changes, rather than cyclic processes, was thus established. However, cyclic processes must still be considered in the analysis, as will be seen in the subsequent lectures.

Planck's Columbia Lectures

THIRD LECTURE: The Atomic Theory of Matter

Planck began the third lecture by defining the role of reversibility and irreversibility for molecular processes. The thermodynamic definition of irreversibility results from the impossibility of certain changes in nature. The principle of conservation of energy results from the impossibility of perpetual motion of the first kind. However, the thermodynamic method is limited when it comes to the impossibility of perpetual motion of the second kind. Planck asked: "Can it not happen that a process, which up to the present has been regarded as irreversible, may be proved, through a new discovery or invention, to be reversible?" He believed that many things would be obtained in future endeavors, which were presently regarded as impossible of attainment.

In the second lecture, the state of a system was defined in terms of energy, pressure, temperature and entropy. Planck argued that the idea of irreversibility is tied to the idea of entropy, and that the problem is the improvement of the definition of entropy, which was first defined by Clausius. Entropy is a function of a reversible process. However, a process of this type cannot be constructed in practice and cannot be measured directly by the physicist.

For example, molecule can be split into other molecular forms or into ions. These molecules can also be re-formed by suitable processes, such as chemical separation methods that involve semi-permeable membranes under the proper conditions. However, not all of the molecules or ions undergo a reversible process. The electrostatic forces that resist molecular separation must be considered as part of this process. The reversible processes are dependent upon forces that are not directly measurable, and theory is therefore dependent upon "thought processes". The irreversible processes, however, allow the successful comparison of the entropy of undissociated molecules with the entropy of dissociated molecules (Lecture II).

The principle of increase of entropy can be evolved from the idea of irreversibility and the impossibility of perpetual motion of the second kind. In resolving the problem of defining entropy, Planck referred to the efforts of Boltzmann, who related the idea of entropy to the idea of probability. Planck referred to the law of the preference of nature and asserted that nature prefers the more probable states to the less probable. Nature processes take place in the direction of greater probability. An example is that heat goes from a body that is at high temperature to one that is at low temperature. The entropy principle is thus reduced to the law of the calculus of probability.

Planck provided evidence of the admissibility of this principle by referring the probability of a physical state back to a variety of possible configurations through which the state may be realized. As an example, he chose the throw of a die with six equal numbered sides, and each side of the die is equally probable to occur in a throw. A throw of two dice that results in total count of two has only one possible "complexion", while the probability of throwing a total count of four is three times as great (the three complexions are 1 and 3, 2 and 2, and 3 and 1).

Planck made a number of assumptions for his atomic theory. All bodies in nature must be constituted of atoms, in accordance with the concept of matter in chemistry. All physical atomic states must be differentiated from one another by a definite reckonable number. He thus introduced the initial concept of quantum states: "...in perfectly continuous systems there exist no reckonable elements --- and hereby the atomistic view is made a fundamental requirement".[11]

The entropy principle is not valid for a single "suspended particle" (atom), because entropy is a function of disordered elements and probable states. The difficulties of measuring the action of a single atom are avoided by measuring only mean values of a system consisting of large numbers of like atoms in disorder. The probability of a state presupposes that all complexions that correspond to a state are equally likely.

With the hypothesis of elementary disorder, it becomes clear that reversible processes are highly unlikely to occur for large numbers of atoms by virtue of the laws of probability. Each atom can be moving in any direction, and with any velocity, that is possible in the three dimensions of space. The particular path and velocity that would reverse this process for any single atom may be indeterminate. A single atom is reversible, while the assemblage of atoms is irreversible. The microscopic observer would be able to see the actions of the atoms, while the macroscopic observer see only the average values of the atoms.

He went further in this assertion, applying it also to heat radiation. His argument is that heat must also possess entropy, since a body that emits radiation experiences a loss

11 Classical analysis is used in the analysis of continuous systems, while the methods of quantum analysis are typically used in atomic physics. However, discontinuous variations, such as step functions and jump functions, which are also quantum variables, can be analyzed using other mathematical methods for linear systems. Therefc re, there are other possibilities, especially classical methods, for analyzing the changes in states of reversible systems.

of heat and a decrease in entropy. The finite amount of heat loss has definite meaning for atomic concepts.[12] The microscopic observer would declare the atomic process to be reversible because all electrodynamic processes can take place in the reverse direction. However, the processes of emission and absorption are not atomic, and the Maxwellian electrodynamic differential equations can retain their validity for discrete elements of heat radiation. The macroscopic observer would define a monochromatic ray through direction, polarization, color and intensity, while the microscopic observer notes all of the irregular variations of amplitude and phase of of the homogeneous heat ray. Heat rays exhibit these types of variations and are lacking frequency coherence.

Entropy is therefore related to disorder and probability, and the next step was to define the mathematical relationships. The probability of two independent configurations having probabilities W_1 and W_2 is

$$W = W1 \cdot W2$$

and the total entropy is

$$S = S_1 + S_2 = k \log W,$$

which applies to both atomic and radiation configurations.

The last step is to define the six general state coordinates. The three space coordinates are φ_1, φ_2, φ_3, and the probability of the state is proportional to the magnitude of the domain, which is a function of the space coordinates

12 Einstein is given credit for the concept of the "particle" nature of light, which conforms with Planck's concept but is more restrictive since it is not defined in terms of size.

and velocities,

$$\int d\varphi_1 d\varphi_2 d\varphi_3 d\dot{\varphi}_1 d\dot{\varphi}_2 d\dot{\varphi}_3 .$$

The velocities in the above equation are a function of time, so it is not an absolute measure of the probability. This time variance is removed by substituting ψ_1, ψ_2, ψ_3, which are defined as the differential ratios of the kinetic potential to the change in state for the three velocities in the above equation,

$$\int d\varphi_1 d\varphi_2 d\varphi_3 d\psi_1 d\psi_2 d\psi_3 .$$

This equation represents the magnitude of the state domain as a function of the position and momentum coordinates. and the probability that the state point falls within the domain.[13] The magnitude of this domain is a measure of the probability that a state point falls within this domain. The above equation for the condition of incompressibility shows that the state domain equation does not change with time. The equations of motion (Hamiltonian) that define the momentum coordinates ψ_1, ψ_2, ψ_3 are functions of energy. Therefore, time has been eliminated from the state equation, and energy concepts have been substituted. This method allows the calculation of the probability of a thermodynamics state for radiant energy, in addition to that of material substances, and these two subjects are covered in the next two lectures.

---

13 This equation can be compared to the *matter wave function* , Ψ, of wave mechanics which quantifies a localized bounding particle.Max Born asserted that the function Ψ^2 is the probability of the separation between the nucleus and the electron of the hydrogen atom.

FOURTH LECTURE: The Equation of State for a Monatomic Gas

In this lecture, Planck derived the equation of state for a material substance: a monatomic gas. This analysis was based on the general laws concerning the concept of irreversibility and the entropy of the gas in a given state. For the equation that defines the entropy,

$$S = k \log W,$$

the essential problem is the calculation of the probability W for a given state of the gas. In order to have a probability and therefore an entropy, there must be a number of possible combinations, and so the state must be macroscopic. A condition of equilibrium cannot be assumed, since this would be special condition in which the entropy is a maximum. Therefore, an unequal distribution of the density may exist in the gas and a variation between the velocities of the molecules. The velocities and coordinates are arbitrary, but certain mean values of densities and velocities must exist.

Every macrostate is associated with a small space in which a large number of molecules have velocities that lie within certain small intervals. The number of the molecules in this macrospace is proportional to

$$dx \cdot dy \cdot dz \cdot d\dot{x} \cdot d\dot{y} \cdot d\dot{z} = \sigma,$$

where σ is termed a "macro-differential" that is proportional to the number of molecules in the macrospace.

The probability factor is dependent upon an arbitrary function and the total number of molecules in the gas,

$$\sum f \cdot \sigma = N, \quad (14)$$

and the proportionality factor, $f(dx \cdot dy \cdot dz \cdot d\dot{x} \cdot d\dot{y} \cdot d\dot{z})$, is a function of the above variables.

The magnitude of an elementary domain is

$$d\varphi_1 \cdot d\varphi_2 \cdot d\varphi_3 \cdot d\psi_1 \cdot d\psi_2 \cdot d\psi_3,$$

wherein the states are defined as

$$\varphi_1 = x, \ \varphi_1 = y, \ \varphi_1 = z, \ \psi_1 = m\dot{x}, \ \psi_2 = \dot{y}, \ \psi_3 = \dot{z}, m = \text{mass}$$

The coordinates and velocities for molecules that lie within each elementary domain σ are within narrow limits, which is fundamental to the calculation of the probability that a molecule lies within this domain. The state domain of a molecule was thus defined in terms of six dimensions; the three dimensions of space and the three dimensional velocities associated with the molecule in this space. The gas is divided into equally probable elementary domains of magnitude $m^3\sigma$.

Planck described an example of N dice that have P uniquely numbered sides that produce probabilities of a throw. The probability, W, of a given state of the molecules corresponds to the number of different kinds of throws of the dice by which it can be realized. In this case, $N = 10$ molecules (dice), and $P = 6$ elementary domains (sides). The unique state chosen is for this example is 3 molecules in the first domain, 4 in the second, 0 in the third, 1 in the fourth, 0 in the fifth, and 2 in the sixth, which can be realized in one instance by a throw resulting in the following numbers on the dice that correspond to each of the six elementary domains:

2, 6, 2, 1, 1, 2, 6, 2, 1 4.

There are 3 dice with the number 1, 4 dice with the number 2, which corresponds to 4 molecules in the second state, 2 dice with the number 6 relates to 2 molecules in the sixth state, etc..

The above throw is but one way to obtain this particular combination of numbers (or molecular state). The total number of combinations by which this set of numbers can be obtained is

$$\frac{10!}{3!4!0!1!0!2!} = 12,600.$$

The primes (!) in the above equation are defined as the factorials $n(n\text{-}1)(n\text{-}2)$---1. Note that the number of allowable combinations was termed the "probability", whereas in today's terminology it is the inverse of this number. There are 10! possible combinations of ten dice with six sides, and there is one chance in 12,600 of getting the above combination of numbers (states) in a single throw. If there were but a single state that predominates for all of the dice, then all ten dice would have the same number for all throws, resulting in only one single combination that is possible. The possibility of getting 10 ones is therefore one in 10!, and the entropy is the maximum value of 3,628,800.

The next steps involve a few mathematical manipulations. The probability function can be expressed in general terms,

$$W = \frac{N!}{\prod (f \cdot \sigma)!},$$

where $\prod$ represents a product series of the terms within the primed parentheses that extend over the P elementary domains.

Substituting into the entropy equation and taking logs,

$$S = k \log N! - k \sum \log(f \cdot \sigma)!,$$

and the summation is over all of the σ elementary domains.

Factorials can be difficult to manipulate mathematically. However, when $(f \cdot \sigma)$ is a large number Stirling's formula can be employed, resulting in the approximation

$$n! \cong \left(\frac{n}{e}\right)^n \sqrt{2\pi n}. \quad (16)$$

A further simplification is obtained by another approximation

$$\log n! = n(\log n - 1),$$

and substituting,

$$S = k \log N! - k \sum f\sigma \left(\log[f \cdot \sigma] - 1\right).$$

Finally, since σ and $N = \sum f\sigma$ are constant for all changes in state,

$$S = \text{const} - k\sum f \cdot \log f \cdot \sigma. \quad (17)$$

Planck thus arrived at a comparatively simple equation for the entropy of gas in terms of the macro-differentials of all of the molecules. The next step was the employment of these concepts to various applications.

In the first application, Planck considered a gas in a state of equilibrium and derived the law of distribution of the velocities of the molecules. The volume of the gas is

$$V = \int dx \cdot dy \cdot dz,$$

and the total energy

$$E = \frac{m}{2}\sum\left(\dot{x}^2 + \dot{y}^2 + \dot{z}^2\right)f\sigma. \quad (18)$$

For a gas in equilibrium, $\delta S = 0$, and from equation (17),

$$\sum(\log f + 1) \cdot \delta f \cdot \sigma = 0, \quad (19)$$

where δf is a change in the law of distribution as a function of N, V and E.

The total number of molecules does not change, so in accordance with equation (14),

$$\sum \delta f \cdot \sigma = 0,$$

and

$$\sum\left(\dot{x}^2+\dot{y}^2+\dot{z}^2\right)\cdot\delta f\cdot\sigma=0,$$

since the energy is constant and through the use of equation (18). The necessary and sufficient conditions for all δf to fulfill equation (19) is

$$\log f+\beta\left(\dot{x}^2+\dot{y}^2+\dot{z}^2\right)=\text{const},$$

or

$$f=\alpha e^{-\beta\left(\dot{x}^2+\dot{y}^2+\dot{z}^2\right)},$$

which is the well known Maxwellian distribution of the velocities, where α and β are constants and the spatial distribution of the molecules is uniform.

The next step was to solve for the constants in the above equation by substitutions of f into equation (14) for the number of molecules in the volume, and into equation (18) for the energy. Then substituting into equation (17), the entropy of the gas is expressed in terms of N, V and E,

$$S=\text{const}+kN\left(\frac{3}{2}\log E+\log V\right). \quad (20)$$

wherein the constant to the right of the equal sign contains terms in N and m.

Applying equation (20) and the general thermodynamic definition of entropy,

$$dS=\frac{dE+pdV}{T},$$

the partial differentials with respect to E and V are easily solved, resulting in

$$p = \frac{kNT}{V} = \frac{RnT}{V},$$

which contains the laws of Boyle, Gay Lussac and Avogadro and R is the absolute gas constant. The constant, k, was then determined by substituting the values for the absolute gas constant, the number of mols of the gas and the mass of a molecule of gas.

From the equation representing the change in entropy as a function of the change in energy, the heat capacities of the gas at constant volume or pressure and the mean kinetic energy L of a molecule was obtained. The mechanical and thermodynamic expressions for the entropy were thus used to derive the above relations.

Planck pointed out that the above methods can be applied to other applications which are less easily handled, such as a monatomic gas that is not in an ideal state. In this case the finite size of atoms and the forces acting among them must be considered (J.D. van der Waals), which involves changes in the probability functions and the energy of the gas. The corresponding change in the condition for equilibrium agrees with that of van der Waals.

Planck took issue with the assumptions made by Boltzmann regarding the more complex system of *polyatomic gases*. The plausible assumption that simple laws of the same kind hold for the intra-molecular process as for the motion of the molecules themselves, results in contradictions. The fact that the intra-molecular energy increases with temperature implies that the molecular heat of a polyatomic gas must be higher than that of a monatomic gas. The application of the Hamiltonian equations of mechanics to intra-molecular motions results in the “law of uniform energy distribution”. Therefore, the mean energy motion of all of the molecules $kT/2$ must be constant in all directions. Applying this principle to both polyatomic gases and the particular case of mercury in a gas state results in a contradiction with measurement, since no heat could then be assigned to intra-molecular energy. The laws of heat radiation provide further contradictions to the law of uniform energy distribution.

Planck ended this lecture by commenting that the principles that have been developed from the kinetic gas theory are not valid for the instable atoms of radioactive substances and this subject is left for future investigations.

The above examples of this lecture were for mechanical systems of gases. The next step is to accomplish a similar result for the radiation process.

FIFTH LECTURE: Heat Radiation, ElectrodynamicTheory

This lecture covers the basic relationships between frequency, energy and spectral radiation and presented a model of the radiating bodies. The basis of the analysis is *electromagnetic theory* of heat radiation, which occurs in the form of light rays. Twenty equations were numbered, and only the most important of these equations will be discussed in this summary.

In determining the energy state of radiant heat, a basic assumption is that energy flowing in a given direction is not affected by heat flowing in another direction. Therefore, the state variables that represent heat radiation are not represented by vectors. However, the energy flowing through a surface element $d\sigma$ and over a time increment dt for the three dimensions of space must be specified in order to determine the state of radiation. The radiation density also varies with the cosine of the radiation angle θ (notice that a different font variable is used for clarity) normal to the surface over the solid angle $d\Omega$ in the direction of radiation. Therefore, the energy flowing through this elementary cone of radiation is

$$K d\sigma\, dt \cdot \cos\theta \cdot d\Omega, \qquad (28)$$

where K is a positive function of space, time and direction. The light generated varies over a wide frequency range, and K is determined from the total radiation over this frequency range,

$$K = 2\int_0^{\infty} \mathfrak{K}_{\nu}\, d\nu, \quad (29)$$

where v is the frequency of unpolarized light and $\mathfrak{R}_v$ is the spectral radiation intensity. Planck (most properly) chose to write this equation in terms of frequency rather than wavelength,

$$\lambda = \frac{q}{v} = \frac{q}{f},$$

where q is the speed of light, which is equal to c in free space (the symbols v and f are used interchangeably in physics and engineering to represent frequency).

The next step was to determine the energy of radiation contained in a unit volume. The equation for the space density of radiation at a point on the surface of a sphere was formed and then integrating over the solid angle from the center of the sphere. Then he obtained the space density of radiation in a cone formed from the center of the sphere and substituted into equation (29),

$$\epsilon = \frac{1}{q} \int K d\Omega = \frac{4\pi K}{q} = \frac{8\pi}{q} \cdot \int_0^\infty \mathfrak{R}_v \, dv. \quad (30)$$

This equation applies under the assumption of uniform radiation intensity, $\mathfrak{R}_v$, and no heat loss in the medium.

The production of heat is emission, while the loss of heat is absorption. Although this process is commonly defined in terms of the surface, both of these processes originate in the atoms or electrons, not at the geometrical surface. Therefore, the surface is a place of entrance or exit of radiation. In determining the thermodynamic state of equilibrium that follows from the second law of thermodynamics, it is assumed that the state of the body varies only with temperature. Therefore, the emission and absorption also vary only with temperature. At far distances from the surface, the influence of the surface is negligible, and heat radiation is in thermodynamic equilibrium everywhere and in all directions. Further, the specific intensity of radiation of a plane polarized ray is independent of the frequency, the azimuth of polarization, the direction of the ray and of the location. The stationary state of radiation is completely determined by the nature of the medium for each temperature.

At thermodynamic equilibrium, a system of bodies, surrounded by a fixed shell impervious to heat, eventually reaches a state in which all of the bodies are at the same temperature. The entropy of such a body has reached its maximum. In a stationary state of radiation, $\mathfrak{R}_\nu$ is determined entirely by the nature of the medium at each temperature. In a medium that is completely diathermanous (transmissive of heat) at a given frequency, there is neither absorption nor emission. Therefore, equilibrium can exist at any finite radiation intensity, since quantum absorption and emission at a given frequency are fundamental to the idea of stationary states. A complete vacuum is diathermanous for arbitrary intensities of radiation and the distribution of the spectral energy.

In the next scenario, he considered the problem of radiation and reflection of two mediums that are separated by a boundary surface. A ray from the first medium is split into a reflected and a transmitted ray at the surface (see Fig. 1). The directions of these two rays vary according to the angle of incidence and the color of the incident ray. The amount of reflected energy is dependent upon the reflection coefficient, ρ, and the transmitted energy is a function of $(1 - \rho)$. In the second medium, the variables are primed values (e.g. ρ'). Invoking the law of refraction,

$$\frac{\sin\theta}{\sin\theta'} = \frac{q}{q'}, \qquad (34)$$

and solving for the geometry of the rays, the resulting equation is

$$\frac{\mathfrak{R}_\nu}{\mathfrak{R}_\nu'} \cdot \frac{q^2}{q'^2} = \frac{1-\rho'}{1-\rho}.$$

For rays polarized at right angles to the plane of incidence that meet at the bounding surface at the angle of polarization,

$$\rho = \rho' = 0,$$

the term on the right of the above equation is equal to 1, and

$$\rho = \rho', \quad q^2 \mathfrak{R}_\nu = q'^2 \mathfrak{R}_\nu'. \qquad (37)$$

The first relation asserts that the coefficient of reflection is the same for both side of the boundary surface (Helmholz), and the second relation that the radiation intensity varies inversely as the square of the velocity of propagation (or as the square of the refractive indices).

In thermodynamic equilibrium, the product

$$q^2 \mathfrak{R}_\nu$$

is a universal function, which may be written as

$$\mathfrak{R}_\nu = \frac{1}{c^2} F(\nu, T) \quad (38)$$

in a complete vacuum. This is a radiation frequency spectrum that also varies with temperature. When the vacuum is contained in a chamber at a uniform temperature, the emission coefficient of the chamber wall has an effect upon the radiation intensity,

$$\rho_\nu \mathfrak{R}_\nu + E_\nu = \mathfrak{R}_\nu,$$

and solving for the radiation intensity produces the ratio

$$\mathfrak{R}_\nu = \frac{E_\nu}{1 - \rho_\nu}. \quad (39)$$

This relationship is the same for all substances (Kirchoff). For perfectly *black radiation*, $\rho_\nu = 0$, and $\mathfrak{R}_\nu = E_\nu$. Black radiation was measured by constructing a chamber having a small hole where the radiation escaped. This experiment was conducted by W. Wien and O. Lummer. Wien constructed a radiation model equation that is fairly accurate at all but the shorter wavelengths.

Planck's goal was to determine the exact form of the frequency spectrum of black radiation. He did not regard this as anything than a very complex problem, and that a common property of the spectral intensity of black radiation is that it depends only upon the temperature. Therefore, the nature of the emitting and absorbing bodies is immaterial. For the emitting bodies, he selected an elementary system, $f(t)$ consisting of an oscillator consisting of two electrical charges that move, with respect to each other, along a straight line along the axis of the charges. The state of the *electron oscillator* is determined by its moment, $f(t)$, which the force that is exerted upon the charges (the electric charge divided by the distance between the charges). The energy of the oscillator is

$$U = \frac{1}{2}Kf^2 + \frac{1}{2}L\dot{f}^2, \quad (40)$$

and the time derivative is the

$$\dot{f}(t) \triangleq \frac{df(t)}{dt}.$$

The constants K and L are the parameters of the oscillator.

For a stable oscillator, $dU = 0$, resulting in the second order differential equation

$$Kf(t) + L\ddot{f}(t) = 0,$$

of which the solution is a pure periodic vibration

$$f = C\cos(2\pi\nu_0 t - \theta),$$

wherein the frequency of oscillation is

$$\nu_0 = \frac{1}{2\pi}\sqrt{\frac{K}{L}}. \quad (41)$$

A plot of this equation for a normalized radian frequency is shown in the following plot:

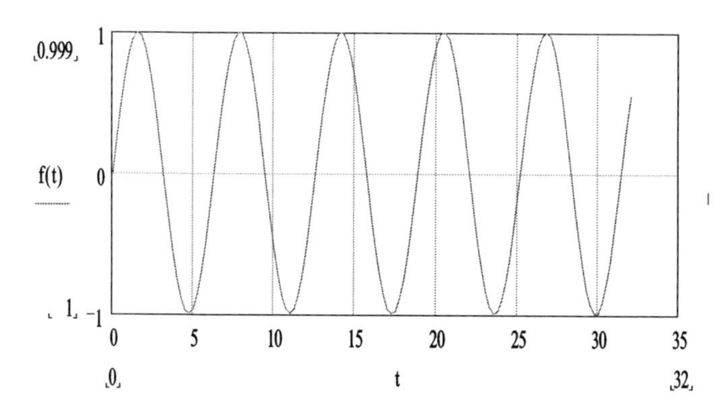

Fig. 9.1 Vibration of an Oscillating Body (Normalized Frequency)

An oscillator, whose energy does not change, neither absorbs nor radiates energy, and therefore the surrounding field could not have any influence on heat radiation. Therefore, the actions from without must be considered.

In accordance with Maxwell's theory of electromagnetic radiation, a vibrating oscillator radiates spherical waves into the surrounding field. In accordance with surrounding field, in accordance with the law of Poynting, would produce a damping effect on the oscillator. However, energy flows both inward and outward, and only a portion of the oscillation energy is lost through radiation. For periodic oscillations, the energy lost is calculated per cycle. Therefore, the energy emitter from the oscillator over the period $\mathfrak{T}$ of the oscillator is represented by the expression

$$\frac{2}{3c^3}\int_t^{t+\mathfrak{T}} \ddot{f}^2(t)dt.$$

If the oscillator is in an electromagnetic field whose electric component is oriented in the direction of the axis of the oscillator, the energy absorbed by the oscillator is

$$\int_t^{t+\mathfrak{T}} \mathfrak{E}_z \dot{f} \cdot dt.$$

Therefore, the following equation represents the principle of conservation of energy:

$$\int_t^{t+\mathfrak{T}} \left(\frac{dU}{dt} + \frac{2}{3c^3} \ddot{f}^2 - \mathfrak{E}_z \dot{f} \right) dt = 0.$$

Under the restriction that the coefficient of the middle term within the bracket of the above integral equation, when multiplied by the radian frequency of the oscillator, is a very small number,

$$Kf + L\ddot{f} - \frac{2}{3c^3} \dddot{f} = \mathfrak{E}_z. \quad (43)$$

This linear differential equation represents the vibrations of the oscillator when excited by an external electric field. The frequency of the oscillator is ν_0, with a small decrement σ.

The next step was to determine the effect when the resonator is in a vacuum filled with stationary black radiation $\mathfrak{R}_\nu$. In general, the electric field can be represented in the form of a Fourier series

$$\mathfrak{E}_z = \sum_{n=1}^{n=\infty} C_n \cos\left(\frac{2\pi n t}{\mathfrak{T}} - \theta_n \right). \quad (44)$$

The Fourier series for the solution of equation (43) may also be determined in a similar manner.

The next step was to calculate the mean energy U of of the resonator using equations (40), (41) and (42). Substituting the value of f from equation (43),

$$U = \frac{3c^3}{64\pi^2 \nu_0^2} \mathfrak{T} \, \overline{C}_{n0}^{\,2}, \quad (45)$$

wherein $\overline{C}_{n0}$ is the mean value of C_n in the series of equation (44).

The intensity of black radiation is obtained from the space density equation (30) and equating it to the average values of the space components of the electromagnetic field. Since the radiation is isotropic, the resulting equation is

$$\epsilon = \frac{8\pi}{c} \int_0^\infty \mathfrak{R}_\nu d\nu = \frac{3}{4\pi} \overline{\mathfrak{E}}_z^{\,2} = \frac{3}{8\pi} \sum_{n=1}^{n=\infty} C_n^{\,2} = \frac{8\pi}{c} \sum_{n=1}^{n=\infty} \mathfrak{R}_\nu \frac{\Delta n}{\mathfrak{T}}.$$

Taking the spectral division of this equation and introducing the mean value and the relationship of equation (45), the result is

$$\mathfrak{R}_{\nu_0} = \frac{3c\mathfrak{T}}{64\pi^2} \cdot \overline{C}_{n0}^{\,2} = \frac{\nu_0^{\,2}}{c^2} U. \quad (47)$$

The many complex mathematical manipulations that Planck utilized in this analysis produced the very simple result expressed in equation (47). The energy spectrum of the oscillators of a system is distorted as a result of the radiation process in which the radiation level is multiplied by the square of frequency. The phenomenon can be illustrated by choosing a system whose *energy* spectrum is flat over the range of frequencies. The *radiation* spectrum for this example appears as in the figure below.

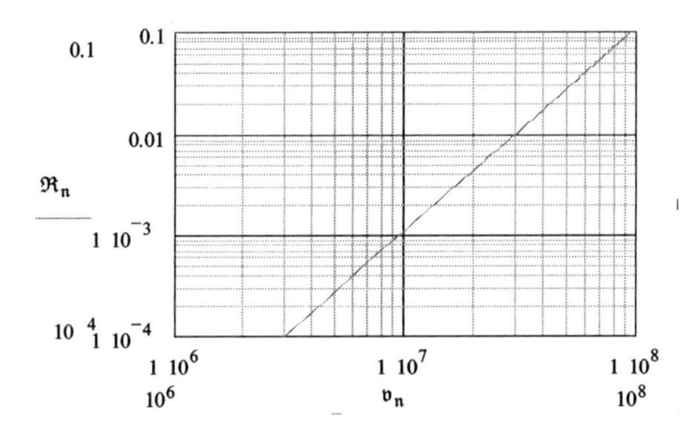

Figure 9.2. Radiation Level of Electromagnetic Oscillators Over a Range of Frequencies

This graph is plotted for oscillators of a unity energy *U*. The shape of this curve correlates with the shape of Wien's model at the low end of the frequency spectrum. Planck was thus able to get the radiation curve to "bend over to the left" (so to speak) by deriving the exact equation that represents this process.

The v_0^2/c^2 (or f_0^2/c^2 with the alternate symbol f for frequency) term of equation (47) appears in many applications in physics and engineering. It appears in the highly accurate antenna radiation equation that is discussed in the next chapter.

The next step is to get the curve to "bend over to the right" at the higher radiation frequencies and to determine the frequency at which the bending occurs.

SIXTH LECTURE: Heat Radiation, Statistical Theory

It is remarkable that Planck was able to derive his radiation equation (53) in only eight short pages at the beginning of this chapter. The derivation is based on just eight simple equations (12, 13,16, 21,40, 47), plus the equations for a state domain and for the probability, *W*, of prior chapters. The insightful nature of Planck's methods become clear as we step through his analytical process.

His first step was to define the energy of a resonator (oscillator) in a black radiation field,

$$\int d\varphi \cdot d\psi ,$$

where φ , which is the state coordinate, and ψ is the state momentum, which is a function of the derivative of φ. Planck's approach was based on energy concepts, and he began by utilizing the energy equation for a resonator

$$U = \frac{1}{2}Kf^2 + \frac{1}{2}L\dot{f}^2. \quad (40)$$

Planck did not define the constants *K* and *L*, saying that the resonator could be either mechanical or electrical. However, he chose the electric dipole with translational motion for his example, as in the fifth lecture. The resonant frequency for his example is $\omega = \sqrt{K / L}$, in accordance with equation (41). For a translational resonant mechanical system, *K* is the spring constant and *L* the mass *m*, while for an electrical circuit 1/*K* is the capacitance *C* and *L* is the inductance. He avoided potential discourse as to the type of resonator by holding strictly to energy concepts in deriving the differential equations for both the stable and irradiated oscillators, although he commented that the same equations can be obtained from electron theory. The final chapter will contain further commentary on this approach.

The next step was to derive the momentum term and substitute it into equation (40). The partial derivative of $U(f, \dot{f})$ with respect to $\dot{f}$ is $L\dot{f} = \psi$, which he calls the "impulse function". After substitution, the resulting state equation is

$$U = \frac{1}{2}K\varphi^2 + \frac{1}{2}\frac{\psi^2}{L}.$$

This is an arithmetic equation that represents an ellipse when plotted, and the area divided by the frequency is equal to the energy of the resonator. Since an ellipse is a closed curve, the energy of a given ellipse is constant for each state. The stable resonator is therefore an oscillator. The ellipse can be represented as a circle by the proper transformation of variables if so desired.

Planck then applied his earlier assertion that the resonators have randomly distributed energies of fixed quantum levels. The difference in energy between any two states is

$$\Delta U = hv,$$

which he calls the "energy element". Each element has equal probability, and the total energy of all of the resonators is the sum of the average energy of the resonators,

$$U_N = NU.$$

According to Planck, the ellipses are "similar and similarly placed", and the differential area between ellipses is constant.

There are a number of steps that he omitted in the above analysis. Important information can be obtained by anlayzing the general equation for an ellipse ,

$$\frac{\varphi^2}{a^2} + \frac{\psi^2}{b^2} = 1.$$

The plot of this equation is shown in Figure 6.1

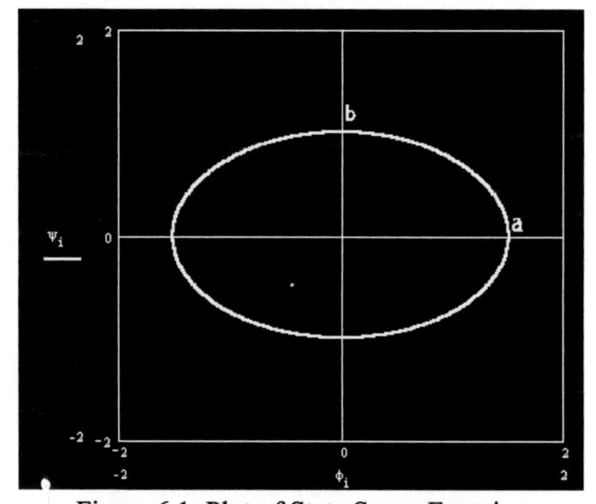

Figure 6.1 Plot of State Space Equation

The values of *a* and *b* are determined by converting the state equation into the form of the ellipse equation and are found to be

$$a = \sqrt{\frac{2U}{K}} \quad b = \sqrt{2LU}.$$

The area of an ellipse is $A = \pi ab$, and upon substitution,

$$A = \pi ab = \pi \sqrt{\frac{4LU^2}{K}} = 2\pi U \sqrt{\frac{L}{K}}.$$

Then substituting this result into the resonant frequency of equation (41),

$$A = 2\pi U \sqrt{\frac{L}{K}} = \frac{2\pi U}{\omega} = \frac{U}{\nu} = \pi ab.$$

Thus the area of the ellipse is equal to the energy of the resonator divided by the radian frequency, and the energy states are related to the parameters of the equations of the ellipses.

For *stable* energy states, the state variables vary cyclically, and the state equation form closed curves, which in this case are ellipses. Planck asserted that there is a definite energy step change that must take place between energy states. The energy difference between each adjacent states must be equal to all others, since they are equally probable. Therefore, each energy element (state) is

$$\epsilon = (U + \Delta U) - \Delta U = \Delta U = \text{constant}.$$

This restriction limits the number of possible states, since the total energy is finite, and there are N states. Therefore, the total energy of the system is

$$U_N = NU.$$

This assertion (by Planck) implies that any stable system must have a finite amount of energy (no singularities).

The shapes of the ellipses vary with the values of the parameters of the equation for the ellipses. Planck avoided the complications involved in the use of these parameters in his computations, utilizing only the area of the curves and the energy state for his analysis. However, additional information may be obtained from a graphical analysis of energy state space, and several different possible scenarios for analysis will be illustrated. In the first case, it will be assumed that the two variables of the state equation vary in direct proportion, as shown in Figure 6.2.

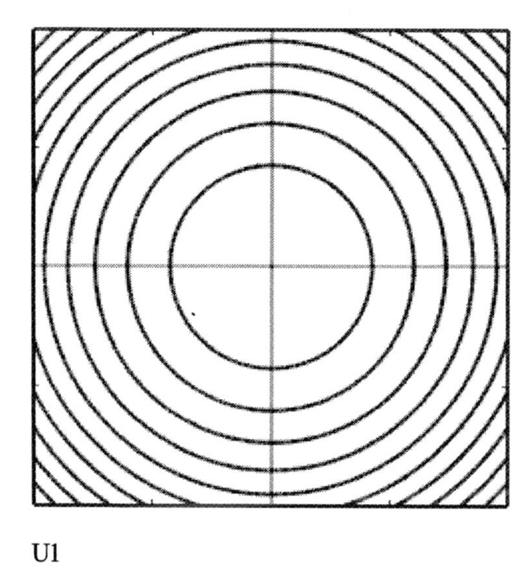

U1

Figure 6.2 Plots of the Energy Equation $U(K,L)$

For each of the above circles, $a = b$, and the "quantum differences" of both state variables decrease as the state energy increases.

As the energy increases further, the number of quantum levels of the state variables increase as shown below.

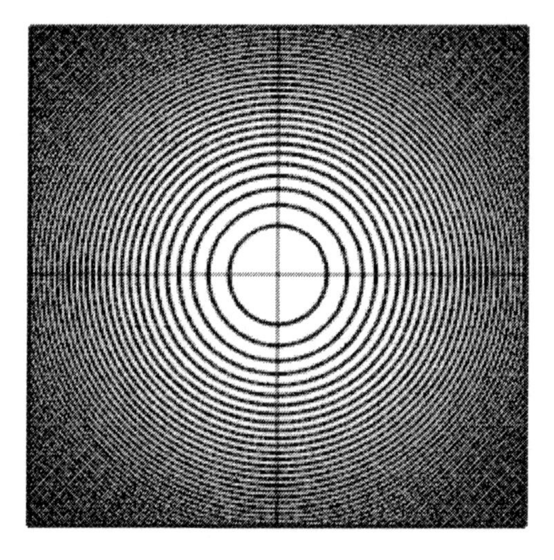

U1

Figure 6.3 Quantum Energy States for Many Energy Levels
With $a = b$ for the Parabola
(The faint patterns are produced by the point-by-point plot method of the computer mathematical program)

Although the areas of the quantum changes are equal to one another, the quantum changes of the two variables are seen to decrease and become more linear as the state energy increases to high levels.

It is more likely that the state variables, φ and ψ , do not vary equally, which is the less restrictive case.

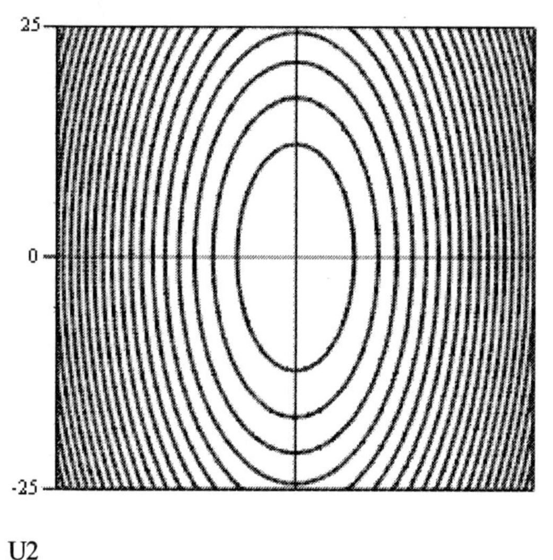

Figure 6.4 Plot of State Equation for Unequal State Variables

In this instance, the state variables, φ and ψ, vary in proportion to each other, which is the simplest case. Whether or not the actual energy states vary in this manner remains to be determined from other implications that follow from other physical restraints. For instance, if the position variable, φ, is limited in its allowable variation, then most of the energy change will take place in the direction of the momentum variable, ψ. The possible variations in the graphical shapes are unlimited from a mathematical standpoint, but not so from a physical standpoint. The vast number of possibilities will not be considered at this point. The plots of the quantum energy states, over a much greater range of energy levels, is shown below.

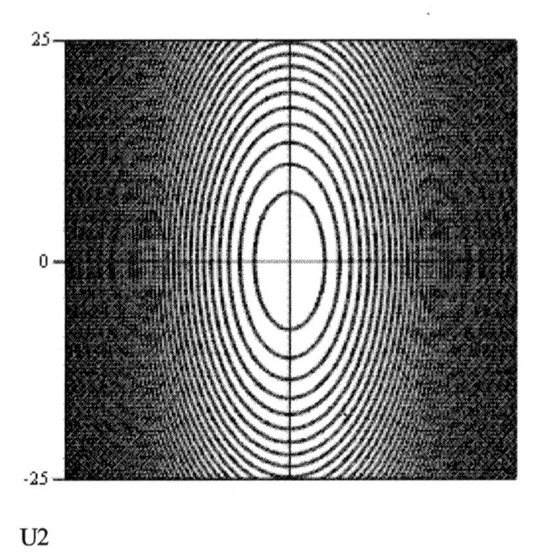

Figure 6.5 Plot of Many Energy States With Unequal State Variables.

In this plot, *b*>*a*, and therefore $\sqrt{KL} > 1$, where *K* and *L* are mechanical parameters. For an electrical system, the equivalent inequality is $\sqrt{L/C} > 1$, which is the inverse of the impedance of the circuit. Therefore, the impedance of the system varies with the eccentricity of the ellipses in the above graph. For the system illustrated above, the impedance is constant for all energy states, since the state variables vary in proportion.

The definition of energy states, as based on the fixed difference in energy between successive states, can be analyzed in different ways. Planck did not utilize graphical analysis, but the above method illustrates the effects of parametric variations of the many energy states under different scenarios. It is also possible that the eccentricity of the ellipses could vary as a function of the system energy, in which case the impedance of the system not constant.

This possibility is less likely, but is possible for propagation through certain substances and is another potential physical effect. As is usually the case, more information is obtained at the expense of additional effort. Other graphical methods will be employed later, in an attempt to better comprehend the implications of Planck's theory.

Planck next employed the concept of entropy, defined in the third lecture, which is a function of probable energy states. Each change in state involves a like energy change, and this energy corresponds to the mean time value for the energy of molecules. The same quantum principle applies to the radiation energy field of the molecules, and this field has a mean space value. In other words, these energies are mean values of the total energy that is distributed among a very large number of resonators. The total energy of the system is equal to the sum of the average energy of all of the resonators within a stationary field of radiation. The total energy is also the sum of the irregular distribution of the individual energies of the resonators, each of which is equally probable. In order to determine the probability distribution of the total energy U_N among the resonators, the first step is to determine the total number of allowable energy states, P. For a fixed energy element, ϵ,

$$U_N = P\epsilon.$$

The probability distribution is a function of the number of elements, N, and the allowable states, P, distributed in all possible "combinations with repetitions". Therefore,

$$W = \frac{(N+P-1)!}{(N-1)!P!},$$

where

$$N! \triangleq 1 \cdot 2 \cdot 3 \cdot 4 \cdot .... (N-1) \cdot N.$$

Note that the number of combinations of the N elements, without the order of the elements taken into consideration, is

$$C_N^P = \binom{N}{P} = \frac{N!}{P!(N-P)!}.$$

The equation for W is obtained by substituting $(N + P - 1)$ for N in the above equation, which allows for the repetitions. Planck called W the "probability", which is a large number. Our present definition of probability means *likelihood*, and is never greater than one, being and inverse function of W.

At this point, Planck employed the entropy equation,

$$S_N = k \log W = k \log \frac{(N+P-1)!}{(N-1)!P!} \cong k \log \frac{(N+P)!}{N!P!},$$

in which the approximation applies for a large number of elements and states. Utilizing Stirling's equation as he did in the fourth lecture,

$$\log N! = N(\log N - 1),$$

and substituting,

$$S_N = k\left[(N+P)\log(N+P) - N\log N - P\log P\right].$$

This is the equation for the entropy of the entire system.

The entropy of a single resonator is

$$S = \frac{S_N}{N} = k\left[(1+\frac{P}{N})\log(1+\frac{P}{N}) - \frac{P}{N}\log\frac{P}{N}\right],$$

wherein the entropy is expressed as a function of the number of resonators and the number of energy states. The total energy is the sum of all of the energies of the resonators. The minimum energy change is constant, and the average energy change is obtained from equation (52), $U_N = P\epsilon = NU$, so

$$\frac{P}{N} = \frac{U}{\epsilon} = \frac{U}{hv}.$$

Upon substitution into the entropy equation,

$$S = \frac{S_N}{N} = S = k\left[(1+\frac{U}{hv})\log(1+\frac{U}{hv}) - \frac{U}{hv}\log\frac{U}{hv}\right],$$

and S is seen to be a function of the mean energy of the resonator.

The final step is to solve the entropy equation for the state energy, U, by applying the relationship between entropy and energy described by equation (49),

$$\frac{dS}{dU} = \frac{1}{T}.$$

The entropy equation can be differentiated directly and solved for U. In order to supply the steps that Planck omitted, a little arithmetic is required,

$$\Rightarrow \frac{1}{kT} = \left[\frac{1}{h\nu}\log(1+\frac{U}{h\nu}) - \frac{1}{h\nu}\log\frac{U}{h\nu}\right]$$

$$\Rightarrow \frac{h\nu}{kT} = \left[\log(1+\frac{U}{h\nu}) - \frac{1}{h\nu}\log\frac{U}{h\nu}\right] = \log\left[\frac{(1+\frac{U}{h\nu})}{\frac{U}{h\nu}}\right]$$

$$\Rightarrow e^{\frac{h\nu}{kT}} = \frac{(1+\frac{U}{h\nu})}{\frac{U}{h\nu}}$$

$$\Rightarrow U = \frac{h\nu}{\left(e^{\frac{h\nu}{kT}} - 1\right)}.$$

This state equation represents the average energy of a resonator in terms of frequency and temperature. It is surprising that Planck did not number the above state energy equation, since it is very important.

The spectral distribution of radiation is different from the spectral distribution of the resonator energy due to the effects of the radiation process. The black radiation energy, $\mathfrak{R}_\nu$, of the surrounding field is “connected” with the resonator energy U

$$\mathfrak{R}_\nu = \frac{\nu^2}{c^2}U. \quad (47)$$

This relationship was determined in a purely electrodynamic manner in the fifth lecture.

There are several basic considerations that were involved in deriving this equation, and it may be helpful to recount the methods that Planck used to establish the foundation of his theory. The interface boundary between the radiator and the absorbing medium affects the radiation on both sides. The surface of the radiating medium can have a different coefficient of reflection, ρ, than that of the absorbing medium. A higher coefficient of reflection at the radiating surface will produce a lower level of radiation, which will therefore also affect the radiation equation. The transmitted energy is a function of the angle of incidence at the surface. Planck invoked the sine law of refraction, which is a function of the ratio of the ray angles on both sides of the interface, and this ratio is equal to the ratio of the velocities of propagation of the two mediums. This function applies to all angles of incidence and polarization, so Planck chose the simplest case, where the plane of incidence is at right angles to the plane of polarization. There is no reflected energy under these conditions, and the result is that the ratio of radiation levels is equal to the inverse of the ratio of velocities of propagation, q and q' of the two mediums,

$$q^2\mathfrak{R}_\nu = q'^2\mathfrak{R}_\nu'^2 = F(\nu,T),$$

where $F(\nu,T)$ is a general function that applies to either medium. In a vacuum, $q = c$, and

$$\mathfrak{R}_\nu = \frac{1}{c^2}F(\nu,T). \quad (38)$$

Substituting the energy state equation into the above equation results in the spectral distribution of radiated energy,

$$\Re_{\nu} = \frac{h\nu^3}{c^2} \cdot \frac{1}{(e^{\frac{h\nu}{kT}} - 1)} . \quad (53)$$

This is the result that Planck sought, which is the equation for black radiation.

As stated earlier, the radiation process produces a different spectral distribution than that of the energy state equation. This effect is illustrated in the graphs illustrated below. Note that these graphs are plotted using logarithmic scales, since the plots of certain functions are in the form of straight lines, thus allowing a clearer picture of the process.

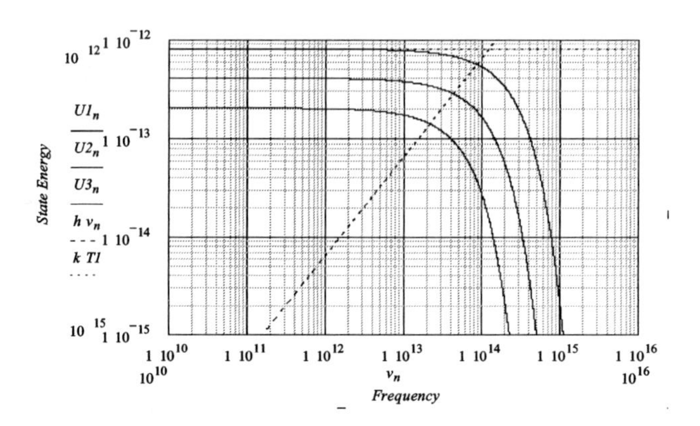

Figure 6.6 . State Energy Spectral Distribution at Three Temperatures (6000, 3000,1500 deg K)

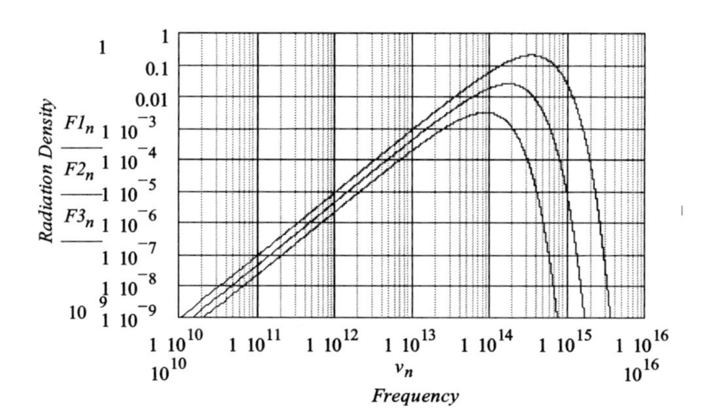

Figure 6.7 . Radiated Energy Spectral Distribution at Three Temperatures (6000,3000, 1500 deg K)

The difference between the spectral distributions clearly illustrates the differences between observed radiation and actual state energy. In comparing the units of measurement in these two figures, it might appear that the radiated energy exceeds the state energy, which is not the case because of the different variables. Radiation occurs in rays, and these rays spread with distance and are spatial functions, whereas the state energy is a function of the movements of the molecules over a very small distance.

In the graph of Figure 6.6, the dotted line is a plot of kT, and the dashed line is a plot of hv. These lines intersect at points on the three temperature-related curves where

$$hf = kT.$$

The intersection of these lines is a one specific frequency, which is a little over 10^{14} Hertz at 6000 degrees Kelvin in the top curve. However, the maximum radiation at this temperature occurs at a frequency that is three times higher in the lower plot of Figure 6.7. The differences between these two plots illustrate the differences between the observed radiation spectrum and the energy state of the system.

In the next chapter, it will be shown that these plots also support Planck's assertion that there is a similar relationship between the behavior of electrons and the "electron gas". At the point where $v^2 = c^2$, the radiation equation (53) reduces to the energy state equation. This is a curious result, since the frequency is equal to the speed of light in a vacuum, regardless of temperature, and the variables have different units of measurement. An answer to this problem is proposed in the next chapter.

The final step in the derivation of his radiation equation is comparatively simple. Optical measurements are most easily obtained by the measurement of wavelength, λ, rather than frequency. Wavelength is an inverse function of frequency, and so the wavelength equation is differentiated in order to obtain an expression in terms of differentials,

$$|dv| = \frac{c \cdot |d\lambda|}{\lambda^2}.$$

Substituting the wavelength differential into the frequency radiation equation,

$$E_\lambda = \frac{c^2}{\lambda^5} \cdot \frac{1}{\left(e^{\frac{ch}{k\lambda T}} - 1\right)}, \quad (54)$$

which is Planck's black radiation equation expressed as a

function of wavelength.

The "secret of minus-1" is embedded in the denominator of equations (53) and (54). Recounting the steps by which it was derived, the probability function W for the number of states was substituted into the entropy equation. An approximation for a large number of states resulted in a simplification of this expression. The derivative of this equation yielded the energy state equation. The use of Stirling's equation, as an approximation, was the key in expressing the entropy equation as a function of the probable states. Electromagnetic theory was used to determine the relationship between the energy of the resonator and the specific intensity of radiation. The effects produced at the interface boundary between the radiating medium and free space was determined, and from this information, the radiation equation, as a function of frequency v was obtained.

Although the radiation equation was converted into a function of wavelength [equation (54)] for ease of measurement, there are important reasons to utilize the frequency dependent equation (53). Fourier analysis can be applied to a frequency spectrum, allowing an estimation of time functions, and this method will be applied to the energy state equation in the next chapter.

Planck next showed that at long wavelengths, his equation reduces to Rayleigh's Law of Radiation,

$$E_\lambda = \frac{ckT}{\lambda^4},$$

while at short wavelengths it reduces to Wien's Displacement Law,

$$E_\lambda = \frac{c^2 h}{\lambda^5} \cdot e^{-\left(\frac{ch}{k\lambda T}\right)},$$

thus confirming the accuracy of his radiation model at both ends of the spectrum.

In the fifth lecture, the space density of a cone of radiation was expressed in the form of an integral equation. This equation was solved by inserting the radiation equation into the integral, resulting in

$$\epsilon = \frac{8\pi}{q} \cdot \int_0^\infty \Re_\nu d\nu = \frac{48\pi h}{c^3} \left(\frac{kT}{h}\right)^4 \cdot \alpha = aT^4,$$

which is the Stefan-Boltzmann Law. Relating the measurements of Kurlbaum to this equation, he was able to solve for the constant,

$$a = 7.061 \cdot 10^{-15} \frac{\text{erg}}{\text{cm}^3 \deg^4}.$$

This equation shows that a quantum change in state energy produces a change in the in the spatial intensity in a cone of black radiation over all space. Note that this result is an apparent contradiction with Einstein's view of light as a particle, since the common definition of a particle has fixed spatial dimensions rather than in the form of a ray which spreads out over space as a function of distance.

The final step in this process was the derivation of the constants, k and h. A third point on the energy curve is needed to make this determination. At the wavelength at which the radiation intensity is a maximum, the change in intensity is zero,

$$\left(\frac{dE_\lambda}{d\lambda}\right)=0.$$

Upon differentiation,

$$e^\beta+\frac{\beta}{5}-1=0,$$

which is a transcendental equation. The solution of this equation is

$$\beta=4.9651=\frac{ch}{k\lambda_m T}.$$

Therefore,

$$\lambda_m T=\frac{ch}{k\beta}=b=\text{const}.$$

which is in accordance with Wien's Displacement Law. Again applying the results of real measurements (by O. Lummer and E Prinsheim),

$b = 0.294$ cm-deg,

resulting in

$h = 6.548 \cdot 10^{-27}$ erg-sec (6.62517 is the correct value),

and

$k = 1.346 \cdot 10^{-16}$ erg/deg (1.38044 is the correct value).

Planck's lengthy methodological approach has finally yielded his famous "Planck's constant" *h*, in addition to Boltzmann's constant.

Upon substitution of Boltzmann's constant into the entropy equation for a physical system,

$$S = 1.346 \cdot 10^{-16} \log W.$$

This equation represents a definitive value for the entropy of a physical system, as defined in earlier lectures, in terms of the probability function W.

Applying Boltzmann's constant to kinetic gas theory results in the ratio of the molecular mass to the mole mass,

$$\infty = \frac{k}{R} = 1.62 \cdot 10^{-24},$$

and

$$N = \frac{1}{\infty} = 6.175 \cdot 10^{23} \text{ molecules/gm-mol}$$

(6.02486 is the correct value).

In the example that he provided, the number of molecules of oxygen in a volume of one cubic cm under ideal conditions was calculated.

He provided other examples of the application of Boltzmann's constant to the determination of physical parameters. The mean kinetic energy of a molecule at one degree Kelvin by substituting the value of Boltzmann's constant into equation (27), from which the mean kinetic energy at any temperature can be determined by multiplying by the temperature. The value of the charge of an electron, in electrostatic units was obtained in a similar manner.

Planck questioned the physical meaning of the constant, h, and said that his theory would be extended by such an extension. In the final chapter of this book, some insight as to the meaning of h will be presented.

At the end of this lecture, Planck discussed the question of whether or not Maxwell's electromagnetic field equations apply *precisely* to the "propagation of electromagnetic waves in a pure vacuum". Other famous scientists of that time believed that radiation occurs in energy quanta, hv, and Einstein went so far as to define these quanta as "particles" that he called "photons". The evidence accumulated over the better part of a century since the time of Planck shows that Maxwell's equations are quite accurate in the analysis and measurement of low-frequency radiation. On the other hand, there is other evidence that tends to support Einstein's photon theory that applies for optical and higher frequencies. This dilemma will also be discussed in the final chapter.

And so this completes the portion of Planck's lectures that apply to the formulation of his blackbody radiation theory. He began with just seven basic equations and used them to derive many of the previously determined laws of chemistry and physics that are based on measurements. He then wove them together the concepts of entropy, thermodynamics and finite energy states. The energy states of a system were based on a model of the atom as an oscillator. Although he claimed that the oscillator could be either mechanical or electrical, he chose the electronic oscillator to develop the model of the atom. The next step was to develop the relationship between state energy and radiation. Then he formulated the probability equation for the energy state of the atom and extended it to the overall probability function for a system consisting of a large number of atoms (or molecules). The entropy equation was then applied and the result equation was solved for the overall state equation that represents the energy of the system. The radiation equation was applied and the result converted from a function of frequency to a function of wavelength to determine his now famous blackbody radiation equation.

The secret of minus-one (the –1 term in the denominator of his blackbody radiation equation) was thus determined from the probability equation for the energy states of the system and the definition of entropy. The method that he used to solve the complex probability equation was the application of Stirling's equation, thus producing a rather simple solution to a difficult problem.

The blackbody radiation equation was "bent over" to the left (in the frequency plot) by virtue of the electromagnetic properties of radiation, and it was "bent over" to the right by the effect of the state probability equation.

In the next lecture, Planck presents his Theory of Least Action. Although this theory is not received much attention in contemporary physics texts, it could be as important as his blackbody radiation theory.

Planck's Columbia Lectures

SEVENTH LECTURE: Planck's Theory of Least Action

The Principle of Least Action applies not only to fields, but to mechanics, thermodynamics and electromagnetics, each of which is a function of *space, time, potential* and *force*, and therefore potentially leading to the more general coverage of a *unified system of theoretical physics.*

The characteristics of reversible and irreversible processes were reviewed at the beginning of this lecture. As Planck put it, irreversibility depends upon the "ensemble of numerous disordered elementary processes of the same kind, each one of which is completely reversible". In an irreversible process, the average or mean value of the elementary process represents the state of the element within a macroscopic region. At equilibrium, each macrostate is equally likely. Irreversible processes are therefore probabilistic, and the entropy equation is a measure of the disorder of a system in terms of the number of energy states and the number of molecules. In probability theory, the properties of the unknowns occur in patterns that can be characterized, resulting in various types of general equations. For instance, the probability equations for the flipping of a coin has only two possible outcomes or states, and its probability distribution is different than the one developed here. The probability equation for W, is basic to irreversible processes of a large number of molecules, and Planck was able to derive the equation for the probability distribution in its exact form. Once the probability distribution was determined, the entropy equation was then complete, leading to the energy state equation and the radiation equation.

It may seem somewhat confusing that all

elementary systems are reversible, while a process that contains a large number of reversible systems can be irreversible. The question is whether or not the actions of a system consisting of a large number of elements can be exactly determined from the actions of the elements. If we knew the response functions of all of the molecules of all different types and all of their initial states, it is conceivable that the entire system could be characterized, and the system would then be considered reversible. However, for a molecular mass, there are a huge numbers of molecules and a great number of energy states to be considered, and this would be an enormous task, even if it were possible to determine all of the initial conditions for all of the molecules. Therefore, it is much more practical to consider it as an irreversible system and utilize probabilistic determinations, as Planck has done, rather than attempting to analyze it as a reversible system consisting of an enormous number of energy states for a huge number of molecules whose initial states are undetermined.

Planck's energy method of analysis has proven to be exceptionally accurate in analyzing the average energy states of irreversible systems. In this lecture, however, Planck pointed out that the principle of conservation of energy does not unambiguously determine the behavior of these processes. He stated that the Principle of Least Action is more comprehensive in making these determinations.

In order to avoid confusion, it is important to note that a large number of equations are used in the development of the Principle of Least Action, and the symbols that Planck used may represent different functions in the various scenarios. In some cases, they are also different from today's common terminology, and these differences will be elucidated as applicable.

The Principle of Least Action, formulated by Helmholtz, is expressed by the integral equation

$$\int_{t_0}^{t_1} (\delta H + A)dt = 0, \quad (57)$$

where δ denotes the displacement of independent coordinates or velocities, A denotes the small increase in energy from external work that the system experiences, and H is the kinetic potential. A system consists of a number of elements and coordinates that may be either finite in number or form a continuous manifold, and Planck uses two sets of examples for these two types of systems.

1. The Position (Configuration) is Determined by a Finite Number of Coordinates.

The first set of examples are for a system having a finite number of elements. In this example, he considers a mechanical system of rigid bodies. For n independent coordinates of position, φ_n, the external work is

$$A = \sum_n \Phi_n \delta\varphi_n = \delta E. \quad (58).$$

In the above equation for the energy increase of the system, δE is written compactly in the form of a summation of terms. The Φ_n are the external force components, $\delta\varphi_n$ are the corresponding differential movements, and E is the energy of the system.

The differential of the kinetic potential H is

$$\delta H = \sum_n \left(\frac{\partial H}{\delta \varphi_n} \delta \varphi_n + \frac{\partial H}{\delta \dot{\varphi}_n} \delta \dot{\varphi}_n \right).$$

Note that there are two components within this equation; one for the variation in position and one for the variation in velocity of each independent coordinate. Substituting the above equation and equation (58) into the integral equation (57),

$$\int_{t_0}^{t_1} dt \cdot \sum_n \left(\frac{\partial H}{\delta \varphi_n} \delta \varphi_n + \frac{\partial H}{\delta \dot{\varphi}_n} \delta \dot{\varphi}_n + \Phi_n \delta \varphi_n \right) = 0.$$

Equating term-by-term,

$$\Phi_n - \frac{d}{dt} \left(\frac{\partial H}{\partial \dot{\varphi}_n} \right) + \frac{\partial H}{\partial \phi_n} = 0. \qquad (59)$$

This expression represents the Lagrangian equations of motion as a function of the changes in position and velocity of the elements of the system. This equation is valid for very small values of $\delta\varphi$ and $\delta\dot{\varphi}$ within the integral. The left hand term Φ_n correlates to the A term (the energy differential) within the integral of equation (57). Planck reduced this equation to a more simple form by first multiplying all terms by $\dot{\varphi}$ and then solving for the energy,

$$E = \sum_n \dot{\varphi}_n \frac{\partial H}{\partial \dot{\varphi}_n} - H. \quad (60)$$

Note that there are only two variables , $\dot{\varphi}_n$ and H, for each term within the summation. These equations of motion and the expression for the total system energy follow from the Principle of Least Action. In the applications that follow, Planck thus applied external work equation (58), the equations of motion (59) and the system energy equation (60) in analyzing the actions of systems.

For mechanical systems, the kinetic potential is

$$H = L - U,$$

where L is the kinetic energy and U is the potential energy. The kinetic energy is a homogeneous function of second degree, and therefore the resulting derivative in the summation term of equation (60) has a coefficient of 2, resulting in

$$E = 2L - H = L + U.$$

Therefore, total energy of the mechanical system is the sum of the kinetic and potential energies, as one would expect from the general laws of physics.

The next example describes a system that consists of a system of linear conductors carrying electric currents. Energy can be introduced in this system through both mechanical actions and electromagnetic action. The equation for mechanical work is introduced in the same manner as was done above,

$$A = \delta E = \sum \Phi_n \delta\varphi_n .$$

Electromotive forces, E_n, are induced in the conductors by electromagnetic fields, as in the case for moving magnets located outside the system. The work done on the system is then

$$W = \sum_n E_n \delta\epsilon_{n\Gamma}$$

Note the change in form of the variables in this equation as compared those used for mechanical systems. The $\delta\epsilon_n$ terms represent the amount of charge passing through the conductors. According to Planck, the charge coordinates ϵ_n cyclical, since the state of a lossless inductor depends only on the time derivatives rather than the instantaneous values. A similar example is the case of mechanical body rotating about an axis, whose state depends upon its rotational velocity.

Note that Planck defined the movement of charges in terms of their *velocities*, rather than the contemporary definition of electrical current that portrays it as simply the number of charges per unit time that are moving through a conductor. This more thorough definition allows the analysis to extend over the regions of mechanical and electrical forces. As was the case for the mechanical system, the kinetic potential consists of two components

$$H = H_\phi + H_\epsilon\,,$$

where H_ϕ is the magnetic potential and H_ϵ is the electrokinetic potential. These potentials are defined in terms of equation (57), which defines the Principle of Least Action.

The electrokinetic potential is

$$H_\epsilon = \sum_{1,n}\left(\frac{1}{2}L_{11}\dot{\epsilon}^2 + L_{1n}\dot{\epsilon}_1\dot{\epsilon}_n\right).$$

The L_{11} term is the self inductance of the first element, and the other L_{1n} terms represent the mutual inductances. The n coordinate positions and velocities are all measured with respect to the position of the first element.

Substituting into equation (59) for the motion of the first conductor,

$$\Phi_1 - \frac{d}{dt}\left(\frac{dH_\phi}{d\dot{\varphi}_1}\right) + \frac{\partial H_\phi}{\partial \varphi_1} + \frac{\partial H_\epsilon}{\partial \varphi_1} = 0,$$

where Φ_1 is the external mechanical force exerted on the first conductor. The electromotive force acting on the first element is a time function of the electric current in the first conductor,

$$E_1 = \frac{d}{dt}\left(\frac{\partial H_\epsilon}{\partial \dot{\epsilon}_1}\right).$$

The relative positions and motions of the other inductive elements with respect to the first element yield a set of similar equations. The force exerted on the first conductor by an external force Φ_1 is determined in this manner.

In addition to the force exerted on the first conductor by Φ_1, there are other mechanical forces produced by the electrical currents,

$$\frac{\partial H}{\partial \varphi_1} = \frac{1}{2}\frac{\partial L_{11}}{\partial \varphi_1}\dot{\epsilon}_1^2 + \sum_2^n \frac{\partial L_{1,n}}{\partial \varphi_1}\dot{\epsilon}_1\dot{\epsilon}_n .$$

The first term to the right of the equal sign is the force acting on the first element that is due to its self inductance, while the forces within the summation of the second term are due to mutual inductances existing between the first element and the other elements. This equation was obtained by partial differentiation of the electrokinetic potential equation shown above. In addition to the external electromotive force E_1, there is another electromotive force, produced by induction, which compensates for the electromagnetic coupled force,

$$-\frac{d}{dt}\left(\frac{dH_\epsilon}{\partial \dot{\epsilon}_1}\right) = -\frac{d}{dt}\left(\sum_n L_{1n}\dot{\epsilon}_n\right).$$

In this case, Planck differentiated with respect to current and time, rather than to the mechanical coordinate variables. The first term within the summation is the inductive voltage drop across the first inductor, L_{11}, produced by the current through it, and the other terms are the voltages produced in the first inductor by the currents in the other inductors.

All of the inductors have resistance, and the current passing through them produces “Joule heat”. This is an irreversible process that is not represented by the Principle of Least Action equation. Helmholtz was able to overcome this difficulty by determining the energy loss as a function of the power dissipated in the resistances of the system elements (the variable w represents resistance) over a time interval. The resulting energy change is

$$dE = -\left(\sum_n w_n \dot{\epsilon}_n^2\right) \cdot dt = -\sum_n w_n \dot{\epsilon}_n d\epsilon_n ,$$

The external force components Φ_n include these additional $-w_n\dot{\epsilon}_n$ (electromotive force) terms.

In the next example, Planck applied the Principle of Least Action to thermodynamic processes. In the prior thermodynamic analysis of energy states, the exact determination of the space coordinates was not necessary, since the variables (V,T, P, S) were not represented directly in terms of position or velocity. The total work exerted on a gas by an infinitesimal displacement is by equation (58),

$$A = -p \cdot \delta V + T \cdot \delta S = \Phi_1 \delta \varphi_1 + \Phi_2 \delta \varphi_2 .$$

In this equation, the first term represents the heat absorbed by the system, and the second term is the mechanical work provided by an external source. The state coordinates are defined as V and S, and their time derivatives are zero for every reversible change of state. Applying the equation of motion (59),

$$-p + \left(\frac{\partial H}{\partial V} \right)_S = 0 \quad \text{and} \quad T + \left(\frac{\partial H}{\partial S} \right)_V = 0,$$

which is a form of the earlier entropy relation

$$dS = \frac{dE + pdV}{T}$$

discussed in the second lecture on thermodynamic processes.

These illustrations show that Helmholtz's Principle of Least Action works produces the same results as the earlier methods for analyzing reversible processes. For irreversible processes, it was shown that additional steps are required in the analysis, and the calculus of probability comes into play. Although this is a difficulty that must be overcome, there is a very important additional factor in this method of analysis that applies to reversible systems. A special term appears when taking the derivative of the energy equation (60),

$$E = \dot{\epsilon}\frac{\partial H}{\partial \dot{\epsilon}} - H .$$

Solving this equation results in the kinetic potential

$$H = -(E - TS).$$

The above equations that were applied to the Principle of Least Action are from thermodynamics, and Helmholtz called the following function,

$$(E - TS),$$

the "free energy" of the system. The first term, E, is the total energy, and the second term is the heat energy. The above equations are in agreement with those of thermodynamics.

As Planck explained it, heat energy is reduced to motion, and yet it is different from mechanical energy "and which in no way can be referred back to it" (the wording of this statement is a bit confusing, since mechanical energy can be related to heat energy by the exchange of energy occurring in the process). This may have been caused by a difficulty in translation, and perhaps a better way to say it is that the process of radiation of energy is quite different from the process that produces a loss of mechanical energy in heat exchange.) Planck rated the importance of reducing heat energy to motion as being on a par with the consideration of light waves as electromagnetic waves.

In section II of this lecture, there are two examples, the first of which is a system of perfectly elastic bodies in which the "generalized coordinates of state form a continuous manifold". In this system, there are six strain coefficients, and the potential energy U is a space integral of a quadratic function f that contains 21 independent constants. Planck's method is the same as in the previous examples, and a term-by-term examination of each equation will not be performed due to the large number of terms. In summation, the result of applying the Principle of Least Action, produces two expressions that define both the conditions within the body and at the surface.

II. The Generalized Coordinates of Lstate Form a Continuous Manifold.

There are two examples in this section. The first example is for a system of perfectly elastic bodies in which the "generalized coordinates of state form a continuous manifold". There are six strain coefficients, and the potential energy U is a space integral of a quadratic function f that contains 21 independent constants, so this will be a bit more complex.

The result of applying the Principle of Least Action, produces two expressions that define both the conditions within the body and at the surface. The coordinates of state are the displacement components, $\mathfrak{v}_x$, $\mathfrak{v}_y$, $\mathfrak{v}_z$, of a material point from its position of equilibrium (x_0, y_0, z_0) as a function of the coordinate variables x, y, z. The external work is given by a surface integral:

$$A = \int d\sigma \,(X_\nu \delta \mathfrak{v}_z + Y_\nu \delta \mathfrak{v}_y + Z_\nu \delta \mathfrak{v}_z)$$

($d\sigma$, surface element; ν, inner normal). The kinetic potential is the difference of the kinetic energy L and the potential energy U:

$$H = L - U.$$

The second expression is for the kinetic energy:

$$L = \int \frac{d\tau\, k}{2}\left(\dot{\mathfrak{v}}_x^{\,2} + \dot{\mathfrak{v}}_y^{\,2} + \dot{\mathfrak{v}}_z^{\,2}\right),$$

wherein $d\tau$ denotes a volume element and k is the volume density. The potential energy U is also a space integral of a homogeneous quadratic function f which specifies the potential energy of a volume element. This depends (geometrically) upon the six "strain coefficients":

$$\frac{\partial \mathfrak{v}_x}{\partial x} = x_x, \quad \frac{\partial \mathfrak{v}_y}{\partial y} = y_y, \quad \frac{\partial \mathfrak{v}_z}{\partial z} = z_z,$$

$$\frac{\partial \mathfrak{v}_y}{\partial z} + \frac{\partial \mathfrak{v}_z}{\partial y} = y_z = z_y, \quad \frac{\partial \mathfrak{v}_z}{\partial x} + \frac{\partial \mathfrak{v}_x}{\partial z} = z_x = x_z,$$

$$\frac{\partial \mathfrak{v}_x}{\partial y} + \frac{\partial \mathfrak{v}_y}{\partial x} = x_y = y_x.$$

In general, the function f contains 21 independent constants that characterize the whole elastic behavior of the substance. For isotropic substances these reduce (on grounds of symmetry) to just two. Substituting the above values into the expression for the principle of least action (57):

$$\int dt \left\{ \begin{array}{l} \int d\tau\, k\left(\dot{\mathfrak{v}}_x \delta \dot{\mathfrak{v}}_x + \cdots\right) - \int d\tau \left(\dfrac{\partial f}{\partial x_x} \delta\, x_x + \dfrac{\partial f}{\partial x_y} + \cdots \right) \\ \qquad\qquad\qquad + \int d\sigma \left(X_y \delta\, \mathfrak{v}_x + \cdots \right) \end{array} \right\}$$

$$= 0.$$

Using the following definitions:

$$-\frac{\partial f}{\partial x_x} = X_x, \qquad -\frac{\partial f}{\partial y_y} = Y_y, \qquad -\frac{\partial f}{\partial z_z} = Z_z,$$

$$-\frac{\partial f}{\partial y_z} = Y_z = Z_y, -\frac{\partial f}{\partial z_x} = Z_x = X_z, -\frac{\partial f}{\partial x_y} = X_y = Y_x,$$

the variations $\delta \dot{\mathfrak{v}}_x, \delta \dot{\mathfrak{v}}_x, \cdots$ and $\delta\, x_x, \delta\, y_x, \cdots$, through suitable partial integration with respect to the variations $\delta\, \mathfrak{v}_x, \delta\, \mathfrak{v}_x, \cdots$, result in the following conditions <u>within the body</u>:

$$k\ddot{\mathfrak{v}}_x + \frac{\partial X_x}{\partial x} + \frac{\partial X_y}{\partial y} + \frac{\partial X_z}{\partial z} = 0, \cdots$$

and <u>at the surface</u> by:

$$X_\nu = X_x \cos \nu x + X_y \cos \nu_y + X_z \cos \nu_z, \cdots,$$

which is known from the theory of elasticity. The quantities $X_x, Y_y, \cdots$ are mechanical surface forces that depend upon the surface conditions.

The second example is the last application of the

Principle of Least Action. It is a special case of electrodynamics, which is an electrodynamic process within a homogeneous isotropic non-conductor at rest (a vacuum). The treatment is similar to the foregoing example. The only difference is that, in electrodynamics, the dependence of the potential energy U upon the generalized coordinate $\mathfrak{v}$ is somewhat different than in elastic phenomena.

The external work is expressed as:

$$A = \int d\sigma \left(X_\nu \delta \mathfrak{v}_x + Y_\nu \delta \mathfrak{v}_y + Z_\nu \delta \mathfrak{v}_z \right), \quad (61)$$

and the kinetic potential is:

$$H = L - U,$$

and therefore,

$$L = \int d\tau \frac{k}{2} \left(\dot{\mathfrak{v}}_x^{\,2} + \dot{\mathfrak{v}}_y^{\,2} + \dot{\mathfrak{v}}_z^{\,2} \right) = \int d\tau \frac{k}{2} (\dot{\mathfrak{v}})^2$$

The energy is expressed by an integral equation:

$$U = \int d\tau \frac{h}{2} (\text{curl } \mathfrak{v})^2 ,$$

and the dynamical equations, including the boundary conditions, are now completely determined. Applying the Principle of Least Action (57):

$$\int dt \begin{Bmatrix} \int d\tau \, k \left(\dot{\mathfrak{v}}_x \delta \dot{\mathfrak{v}}_x + \cdots \right) - \int d\tau \, h \left(\text{curl}_x \mathfrak{v} \delta \, \text{curl}_x \mathfrak{v} + \cdots \right) \\ + \int d\sigma \left(X_\nu \delta \mathfrak{v}_x + \cdots \right) \end{Bmatrix}$$

Using the methods employed above for the theory of elasticity, within the interior of the non-conductor:

$$k\dot{\mathfrak{v}}_x = h\left(\frac{\partial \operatorname{curl}_y \mathfrak{v}}{\partial z} - \frac{\partial \operatorname{curl}_z \mathfrak{v}}{\partial z}\right), \cdots$$

which can be written more simply as:

$$k\ddot{\mathfrak{v}} = -\, h \operatorname{curl} \operatorname{curl} \mathfrak{v}, \qquad (62)$$

and at the surface:

$$X_v = h\left(\operatorname{curl}_z \mathfrak{v} \cdot \cos vy - \operatorname{curl}_y \mathfrak{v} \cdot \cos \ldots vz\right), \cdots \quad (63)$$

If we identify L with the electric, and U with the magnetic energy,these equations are identical with the known electrodynamical equations. If we put

$$L = \frac{1}{8\pi}\int d\tau \cdot \epsilon \mathfrak{E}^2 \quad \text{and} \quad U = \frac{1}{8\pi}\int d\tau \cdot \mu\, \mathfrak{H}^2,$$

($\mathfrak{E}$ and $\mathfrak{H}$ are the field strengths, ϵ is the dielectric constant, and μ i s the permeability for this expression) and compare these values with the above expressions for L and U:

$$\dot{\mathfrak{v}} = -\,\mathfrak{E} \cdot \sqrt{\frac{\epsilon}{4\pi k}}, \quad \operatorname{curl} \mathfrak{H} \cdot \sqrt{\frac{\mu}{4\pi h}}. \quad (64)$$

It follows then, by elimination of $\mathfrak{v}$, that:

$$\dot{\mathfrak{H}} = -\sqrt{\frac{\epsilon h}{\mu k}} \text{ curl } \mathfrak{E}.$$

By substitution of $\dot{\mathfrak{v}}$ and curl $\mathfrak{v}$ into equation (62) (found above for the interior of the non-conductor):

$$\dot{\mathfrak{E}} = -\sqrt{\frac{\mu h}{\epsilon k}} \text{ curl } \mathfrak{H}.$$

With the known electrodynamical equations expressed in Gaussian units:

$$\mu\dot{\mathfrak{H}} = -c \text{ curl } \mathfrak{E}, \quad \epsilon\dot{\mathfrak{E}} = -c \text{ curl } \mathfrak{H}$$

(c is the velocity of light in a vacuum) results in a complete agreement, if we put:

$$\frac{c}{\mu} = -\sqrt{\frac{\epsilon h}{\mu k}} \text{ and } \frac{c}{\epsilon} = -\sqrt{\frac{\mu h}{\epsilon k}}.$$

From either of these two equations it follows that:

$$\frac{h}{k} = \frac{c^2}{\epsilon\mu},$$

the square of the velocity of propagation.

Utilizing equation (61), the energy entering the system from without is:

$$dt \cdot \int d\sigma \left(X_\nu \dot{\mathfrak{v}}_x + Y_\nu \dot{\mathfrak{v}}_y + Z_\nu \dot{\mathfrak{v}}_z \right),$$

or by surface equation (63):

$$dt \cdot \int d\sigma\, h\left\{\left(\mathrm{curl}_z\, \mathfrak{v} \cos vy - \mathrm{curl}_y\, \mathfrak{v} \cos vz\right)\dot{\mathfrak{v}}_x + \cdots\right\},$$

from which the substitution of the values of $\dot{\mathfrak{v}}$ and curl $\mathfrak{v}$ from (64) turns out to be identical with the Poynting energy current.

By the application of the principle of least action, with a suitably chosen expression for the kinetic potential *H,* Planck arrived at the Maxwellian field equations.

Recounting the steps in this last example, the Principle of Least Action was applied to the electrodynamics of an isotropic medium, which in this case is a vacuum. Planck then applied a method that is similar to that used for the theory of elasticity. After deriving the proper expression for the kinetic potential *H*, he derived Maxwell's volumetric field equations and the Poynting energy, which is a surface function.

In electromagnetic analysis, the displacement components are space functions, as was the case for mechanical systems. However, Planck pointed out that it is not generally possible to interpret the displacement variable $\mathfrak{v}$ as a mechanical quantity for electromagnetic systems. The reason for this difference is that the rotation of the field is curl $\mathfrak{v}$, and therefore the velocity $\dot{\mathfrak{v}}$ is constant for an electrostatic field. With a constant velocity, the displacement increases constantly with time, and therefore "curl $\mathfrak{v}$ can no longer signify a rotation". There is a problem with the interpretation of the spatial characteristics of electrical phenomena as mechanical forms (electrical phenomena are "perfectly elastic"), and Planck cited a reference regarding this difficulty. However, he maintained that the Principle of Least Action can be extended beyond ordinary mechanics, and that this principle can be utilized as the foundation for general dynamics, since "it governs all known reversible processes".

The material in this chapter was a precursor of electrical circuit theory and finite element analysis of mechanical systems. In fact, it extends beyond these modern methods, in certain respects, in that all possible conditions are considered. Also, Planck defined the variables of electrical systems in terms of electric charge, rather than current, and included the space variables (which, for instance, is not present in the definition of electrical current).

In the next chapter, a special hypothesis, the principle of relativity, establishes the dependence of the kinetic potential upon the generalized coordinates and their derivatives (velocities). In the final chapter, I will present an alternate method of analysis that avoids this difficulty and allows a proper definition for the function curl $\mathfrak{v}$. The differences between the two methods and advantages and disadvantages will also be discussed.

EIGHTH LECTURE: The Principle of Relativity

Planck has presented a very clear and concise description of the General Theory of Relativity in this last lecture. There are some interesting points to be made regarding the history of the development of this theory and of the key assumptions upon which it is based.

An early hypothesis that preceded the theory of relativity was presented by physicist G. A. Fitzgerald, who argued that there is an apparent contraction in size of objects occurs that varies as a function of v^2/c^2 due to the presence of the "quiescent ether" in the universe. The assumption of an ether that is similar to a material medium was shown to be contradictory by the Fizeau experiment cited by Planck. H. A. Lorentz formulated a set of equations, based on the Fitzgerald contraction, as a function of time and space. This mathematical model allowed a resolution of the contradiction between theory and measurement, but the meaning of the equations and their parameters was up to question.

Although the Lorentz equation was based on the presence of an ether, the general relativity theory of A. Einstein was based on the Lorentz equation does not depend upon the presence of an ether but is based on a different assumption. Einstein's conceptual description was framed into a mathematical form by H Minkowski. Planck discussed the differences between the two theories. The Fitzgerald/Lorentz theory assumes that a *quiescent ether* is the *carrier* of electromagnetic waves. The general assumption had been that the velocity of a ponderable body can be measured with respect to the ether. With the relativity principle, the dependence upon the presence of an ether drops out, along with the mechanical explanation of electrodynamic processes and the assumption that space contracts with velocity. Another problem with the ether theory is that electrodynamic processes are continuous, while mechanical motions are discrete, as was illustrated in the prior lecture. Planck pointed out that if electromagnetic process were to occur in a ponderable medium (ether), then *it would not then be possible to distinguish between field strength and induction*. For the relativity theory, which assumes a complete vacuum in which electromagnetic energy is propagated like ponderable atoms, no physical properties can be assigned to the vacuum. The velocity of propagation is not a property of the vacuum, and "where there is no energy there can be no velocity of propagation". The velocity of propagation is only dependent upon electromagnetic energy and not the medium.

In the Einstein/Minkowski model, time is allowed to vary with respect to two observers, one of which is moving with respect to the other. The Lorentz equations were modified to produce an apparent variation in the observed dimensions of spatial objects. This last assumption is the most radical. As Planck emphasized, "this new conception of time makes the most serious demands upon the capacity of abstraction and the projective power of the physicist". While Planck provided several examples of physical phenomena that are in accordance with this concept.

Another assumption applies to electromagnetic radiation, which is believed to travel out as spherical waves (this assumption is also basic to the application of Maxwell's field equations). The spherical wave equation is

$$x^2 + y^2 + z^2 - c^2t^2 = 0. \quad (66)$$

For the second observer of the wave, x, y, z and t vary, resulting in the modified wave equation,

$$x'^2 + y'^2 + z'^2 - c^2t'^2 = 0. \quad (67)$$

Only the *relative* position of one to the other affects the observation. The Lorentz equations of transformation describe the change in dimensions of the variables in the wave equation. For the first observer,

$$x' = \frac{c(x - vt)}{\sqrt{c^2 - v^2}}, \quad t' = \frac{c(t - \frac{vx}{c^2})}{\sqrt{c^2 - v^2}}, \quad (68)$$

while for the other observer,

$$x = \frac{c(x' - vt')}{\sqrt{c^2 - v^2}}, \quad t = \frac{c(t' - \frac{vx'}{c^2})}{\sqrt{c^2 - v^2}}. \quad (69)$$

In the example that was illustrated in Figure 5, only relative motion in the x-direction affects its apparent dimension, and the other dimensions remain unchanged as is shown in these equations.

The equations of transformation, as developed by Minkowski for a four-dimensional system (x, y, z, ict), correspond to a rotation through the imaginary angle iv/c. This mathematical model describes the novel concept of the fourth dimension of time, which has received much publicity in the general public. Uncharacteristically, Planck did not offer any physical example in which the imaginary angle appears. The question of the physical concept of rotation with respect to this imaginary angle will be discussed in the final chapter, along with an example.

Planck applied the relativity theory to the Fizeau experiment in which the difference between the velocities of the waves in the upper and lower tubes results in a change in velocity in lower tube with respect to that of the to the top tube. The resulting equation (the Fresnel coefficient) is

$$2v\left(1-\frac{q_0^{\,2}}{c^2}\right)=2v\left(1-\frac{1}{n^2}\right),$$

in which n is the index of refraction that is a measure of the speed of light in a medium. This result is in agreement with the measurements of Fizeau, thus providing a resolution of the contradiction mentioned above.

While the relativity principle extends to optical, electrodynamic and ordinary mechanics phenomena, Planck observed that the kinetic energy of a mass point

$$\frac{mq^2}{2}$$

is "incompatible with this principle". The meaning of this statement was described by applying the Principle of Least Action.

According to Planck, the Principle of Least Action applies to all physical phenomena. The kinetic potential H of a space element in a physical system varies according to the law

$$H\cdot dt=H'dt'$$

with respect to the two observers in the above physical example. Therefore the following relationship also applies,

$$\frac{H}{\sqrt{c^2 - q^2}} = \frac{H'}{\sqrt{c^2 - q'^2}},$$

and

$$H = \sqrt{1 - \frac{q^2}{c^2}} \cdot H_0 .$$

The Lagrangian equations of motion (59) apply, and in accordance with equation (60) the kinetic energy of the mass is found to be

$$E = -\frac{H_0}{\sqrt{1 - \frac{q^2}{c^2}}}$$

and the momentum

$$G = -\frac{qH_0}{c\sqrt{c^2 - q^2}}.$$

The "transverse mass" is defined as G/q, and the "longitudinal mass" as dG/dq. For a system at rest (fixed positions of one observer to the other), $q = 0$, and for $H_0 = m_0$

$$m_t = m_l = m_0 = -\frac{H_0}{c^2} = -\frac{m_0}{c_0}.$$

Substituting into the above equations, the transverse mass is

$$m_t = \frac{m_0}{\sqrt{1-\frac{q^2}{c^2}}},$$

and the longitudinal mass is

$$m_l = \frac{m_0}{\left(1-\frac{q^2}{c^2}\right)^{\frac{3}{2}}}.$$

The two expressions are equal for $q = 0$ and both increase with velocity, but the longitudinal mass becomes predominate as q approaches the speed of light.

The kinetic energy is

$$E = \frac{m_0 c^2}{\sqrt{1-\frac{q^2}{c^2}}},$$

which illustrates the variation from the value of ordinary mechanics

$$E_0 = \frac{m_0 c^2}{2}$$

previously described. Therefore, in the relativity theory both the energy and the mass increase with the velocity of the moving body.

The final example is the case where the body is in

the form of black body radiation. Planck claimed that, for quasi-stationary processes, it is the only process whose dynamics are known with absolute accuracy. For black body radiation at rest, the energy is described by the Stefan-Boltzmann law

$$E_0 = aT^4V,$$

the entropy is

$$S_0 = \int\left(\frac{dE_0}{T}\right) = \frac{4}{3}aT^2V,$$

and the pressure is

$$p_0 = \frac{aT^4}{3}.$$

Therefore, the kinetic potential of a stationary body is

$$H_0 = \frac{aT^4V}{3},$$

and for moving black body radiation, the kinetic potential is

$$H = \frac{aT^4V}{3\left(1 - \frac{q^2}{c^2}\right)^2}.$$

The thermodynamic quantities, pressure, energy, momentum, and longitudinal and transverse masses, of the moving black body radiation are uniquely determined from this equation. Because of this property, the kinetic potential was called an “invariant”.

In this theory, a moving body has two masses: a transverse mass m_t and a longitudinal mass m_l. This would suggest the possibility that mass is a vector quantity!

In his closing statements, Planck clearly regarded the Principle of Least Action and the invariants in the system of physics, such as those described above, as being among the most important advances in the state of physics, since they will retain their meaning for all investigators for all time. Future efforts were to be directed toward lasting results (“so to speak, for eternity”).

Planck's Columbia Lectures

CLOSING COMMENTS ON PLANCK'S LECTURES:

In each of these lectures, the simplicity and clarity of Planck's presentations stand out. The subjects of his studies, however, were extremely complex, which is even more outstanding. He did admit that the processes involved in deriving these theories were quite involved and required a great deal of effort, including those of many other scientists. It is notable and honorable that he gave credit to many of these scientists for their contributions, especially to Boltzmann, who devoted some 40 years of effort in developing the theories of thermodynamics.

The information provided in these lectures would, I believe, be a good source of material for teaching. For instance, the first two lectures contain information pertinent to a course on thermodynamics and chemistry. The lecture on radiation is applicable to courses in engineering and physics. I was surprised to find that the material in the seventh lecture encompasses a broad range of subjects that covers mechanical structure analysis, acoustic waves, etc., with certain similarities to various modern methods. The final lecture on the theory of relativity provides a concise and clear description of the various relationships of time, space, and energy as depicted by Einstein.

I hope that I have been able to clearly and accurately represent Professor Planck's presentation. The next chapter is entirely mine. In it, I present a new and somewhat different view of the system of physics. For example, the definition of electric current is modified. Is the current in a wire the same at all points along its length under all circumstances? It is my belief that the coordinates of space should be included in this definition, and I was surprised to find that Planck had utilized a similar approach in his lecture on The Principle of Least Action. This change in definition allows a much greater range of possibilities and leads to some interesting results, as will be shown.

Professor Planck stated that mathematics is a tool for physicists, engineers, etc., and that excessive use of abstractions is "dangerous". The abstract methods of quantum mechanics seem to fit in this category. The methods of Planck, as illustrated in this book, are far superior to any scientific method that is unclear, especially for those of experts in their field. Students of science that take the time and effort to learn and apply such methods in their future efforts can profit. Planck's methods are thorough, clear, detailed but concise, based on reality, provide greater insight and visualization and lead to successful and accurate results.

For those who decide to read the next chapter, there will be some surprises. Some important existing theories, that may be based on faulty or incomplete assumptions, will be challenged, and the results that are obtained were not as had originally been expected.

Planck's Columbia Lectures

Chapter 10

New Ideas and Concepts

Introduction

This chapter is different from the prior chapters, which were devoted entirely to Planck's lectures. Some new physical models will be discussed, and evidence to support the conformance with measurements and Planck's theories will be presented. The thoughts and ideas presented here are my own, some of which have been expressed in my articles in technical journals and my books (see references).

There are three sections in this chapter. The first section covers some of the important developments of electron theory over the past century and integrates the methods of analysis with those of Planck.

Planck's Columbia Lectures

Electron Theory:

Several different concepts of electron theory will be threaded together in this section, including Planck's blackbody radiation model.

Planck referred to atoms as being "oscillators", and he chose to use the electric oscillator to model the basic atom[1]. This model provides some support for the fundamental assertion, made in my earlier books, that all matter is electrical in nature. I believe that electron theory is basic to the study of the characteristics of matter and that the analytical methods of electronics will become fundamental to the physics of matter in the future. Certain accomplishments of other scientists in the development of electron theory are reviewed and shown to correlate, in various ways, with Planck's theory.

According to Planck, "All physical ideas depend upon measurements…". Michael Faraday, who made significant contributions to the characterization of electric fields, exemplified this principle. Another scientist, who also believed strongly in the importance of measurements and made important scientific discoveries, was not properly recognized for the significance of his accomplishments. Oliver Heaviside was born in London, England in the year 1850. His background is similar in some ways with that of Faraday, having had no schooling past the age of 16. However, he did not have the advantage of being directly under the guidance of a top scientist, and so he trained himself at home in languages, mathematics and the natural

---

1 Planck also stated that the model of the atom could just as well be mechanical in his sixth lecture. The analogy between electrical and mechanical models is well known. The dynamics of electrical and mechanical systems has been my primary subject of study for some many years.

sciences. He became a telegraph operator in 1868, but retired in 1874 due to deafness and began to experiment with telegraphy signal transmission. Using unique mathematical methods, he developed the basic methods of vector analysis that are used today.

Heaviside also made contributions in the application of Maxwell's equations, which are now commonly written in Heaviside's simplified vector form (four equations in two variables), and electromagnetic theory, and he published an important set of papers on the subject from 1892 to 1912. In 1901, Marconi first sent radio signals around the earth, but no one could explain this phenomenon. Previous measurements showed that electromagnetic rays travel in straight lines, while Marconi's signals must have somehow followed the curvature of the earth. After a year of study, Heaviside concluded that the radio waves were trapped in a layer in the atmosphere. In his words, waves traveling around the earth "might accommodate themselves to the surface of the sea in the same way as waves follow wires" and "there may possibly be a sufficiently conducting ionized layer in the upper air". His hypothesis has been proven to be correct, and this atmospheric layer is now known as the Heaviside or E layer.

Heaviside coined the words *impedance*, *inductance* and *attenuation*, and applied these concepts in the analysis of electrical circuit operation, including transmission line theory. However, his most important contribution was the introduction of a new method of circuit analysis utilizing unique methods of operational calculus. He railed against the exclusive use of rigorous (Euclidean) mathematical methods in scientific theory and took liberties in developing new analytical methods. The beauty of his calculus method is that the differential equations that represent the model of an electrical circuit can be solved by simple algebraic methods. The resulting algebraic equations are generally easily solved and then transformed back into an equation that represents the solution of the differential equation by using a conversion table.

Although highly successful, his method was not accepted by mathematicians, who objected to its lack of rigor. Heaviside argued that mathematics is simply a tool, and that real measurements should be the primary defining factor. Eventually, the method of Heaviside eventually gained recognition, but only after acceptable mathematical methods for his method were developed. Several famous mathematicians and physicists contributed to the further development of Heaviside's operational calculus. Cauchy provided a mathematical validation of Heaviside calculus, which is often referred to as Cauchy-Heaviside calculus. The architects of *quantum mechanics*, Dirac, Born, Heisenberg, Hilbert, Von Neumann and Wiener, utilized and extended Heaviside's operator method, applying it to atom theory.

Heaviside's methods of analyzing electrical circuits are now commonly used in engineering, and a brief description of his methods is presented in order to help illustrate a certain commonality with Planck's theory. In laboratory experiments of the responses of electrical systems to various inputs, he discovered that an electrical circuit that is excited by a very sharp input signal always produces the same response that characterizes the system. He subsequently defined the *unit step* function, or *Heaviside step function*, as a step change from one energy or voltage level to another that occurs in zero time. The *unit impulse function* $\delta(t)$ is a pulse of energy that occurs in zero time. The response of a linear system to a unit impulse produces a unique characteristic time response $h(t)$, while the response to a unit step function produces a time response that is the integral of the characteristic response. Either one of these signals can be used to predict how the system will react to a known input signal. Note that such functions are, in fact, represent quantum changes.

To illustrate the Heaviside technique, consider a simple example of a simple electrical circuit having an input signal to the system that produces an output signal. The characteristic response to an impulse is $h(t)$, and its Fourier transform is $H(\omega)$, where ω is defined as the radian frequency. For an arbitrary input signal $f(t)$, the Fourier transform of $f(t)$ is $F(\omega)$. The resulting output of the system is simply

$$G(\omega) = H(\omega) \cdot F(\omega). \quad (10.1)$$

The output response time function $g(t)$ can be found from a table of transforms or can be calculated using the inverse Fourier transform. The function $H(\omega)$ is generally called the frequency response of the system.

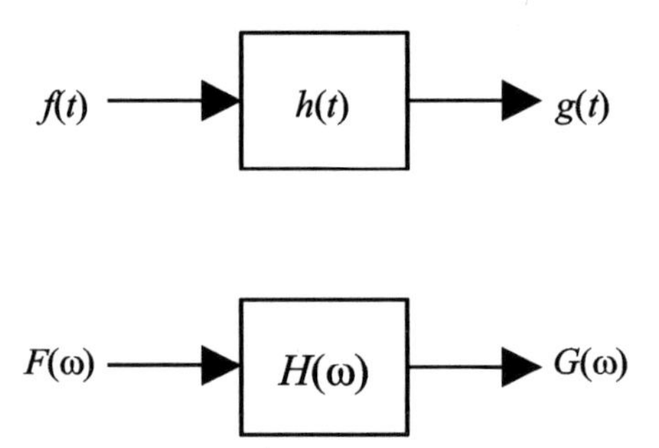

Figure 10.1. Time and Frequency Functions of a Network

A unit pulse (time function) is pictured below:

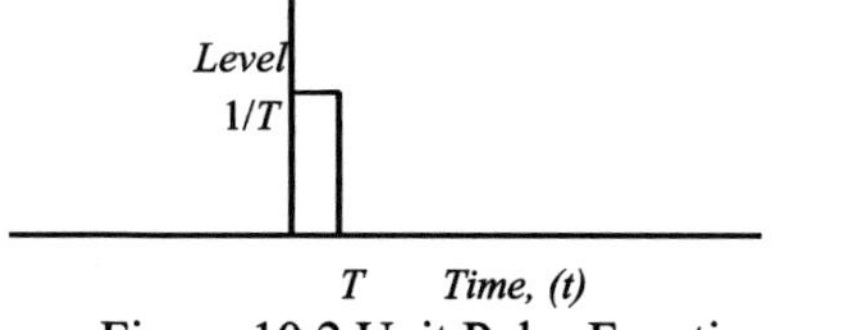

Figure 10.2 Unit Pulse Function vs Time

The frequency spectrum of this pulse is a (sin $a\omega t$)/$a\omega$ function, and the shorter the time pulse the wider the frequency spectrum ($a = T/2$). As the width of the pulse T is decreased, the peak level increases such that the area of the pulse is unity. This process is carried on indefinitely, resulting in the definition of a *unit impulse*, whose spectrum is flat and extends out indefinitely, as shown in Figure 10.3.

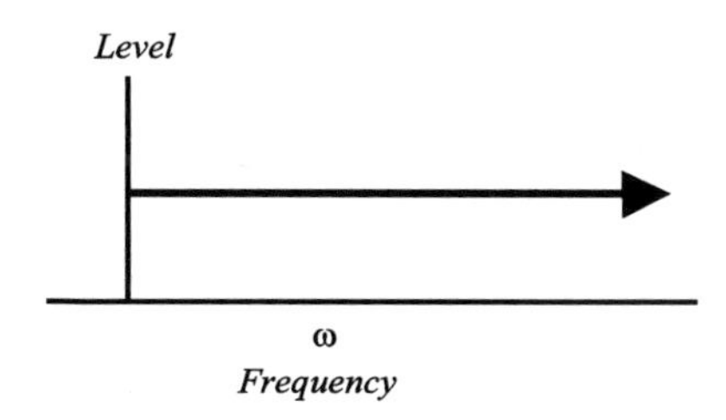

Figure 10.3. Frequency Spectrum of a Unit Impulse

A function of this type falls within the domain of mathematical *singularities*, since the peak energy of a step function approaches infinity with time. In order to resolve the infinite energy problem the level could be reduced, but if the level were zero there would not be an input signal. Singularities present these types of problems in mathematics, which can be very difficult to resolve. In this case the problem remained unsolved for many years, and mathematicians of the time chose to discount Heaviside's method. In the resulting controversy, Heaviside argued strongly against the pure rigorous approach in the application of mathematics, arguing that "…mathematics is an experimental science, and definitions do not come first, but later on." It is the responsibility of the mathematicians to find methods to make mathematics conform to measurements.

A mathematician (Laurent Schwartz) developed an acceptable mathematic method of handling the impulse function in 1950. A major difficulty with the rigorous mathematical method is that the inverse integral transform of the impulse function appears to have no meaning (Papoulis). The reason for this conclusion involves the circumstances of the mathematical process. Expressing the limiting process by an integral equation that contains an impulse function that is not in the form of an "ordinary function", does not necessarily uniquely specify the unit impulse function. This mathematical method involves representing the impulse function as the limiting case of a distribution. The use of distributions, however, leads to other mathematical problems that are quite involved, and a number of additional mathematical theorems are required to make the process rigorous.

Heaviside's method works so well that has now been uniformly adopted and used in various types of engineering efforts. It would, however, be advantageous to have an impulse function that can be defined in a way that can be clearly defined in a mathematical sense that will eliminate the problems associated with the mathematical limiting process. One way to resolve this problem is to limit the bandwidth of the spectrum of the impulse in a way that would not affect the accuracy of the analysis, since this would limit the both the peak and average energies of the impulse and its derivatives, thus providing a clear and realizable impulse time function. The problem is to find a real function that has the proper spectral characteristics. The most straightforward method is to simply exclude all frequencies above a certain specific very high frequency. However, systems of this type do not occur in nature. All known realizable systems have a time response that gradually falls off with time and frequency. It will be shown that some of Planck's equations can be utilized to provide a suitable answer to this problem. The first step is to establish the link between Planck's theory and a corresponding impulse function.

The proper impulse function must be measurable, and it must have a flat frequency response over a wide range of frequencies. One type of signal that has this characteristic is the *random noise* that is always present in electrical circuits. Although the levels of electrical noise can be extremely low, these signals are easily detected and measured. We now know a great deal about the characteristics of this type of noise, which will now be reviewed to see if an applicable noise impulse can be characterized and whether or not it fits the requirements of a proper impulse function.

The characterization of the ambient noise signals of electrical conductors was first accomplished by electrical engineers in 1927-1928. J. B. Johnson of Bell Telephone Laboratories conducted a set of experiments on a variety of electrical conductors including resistors, chemical compounds and solutions. Using instruments that could measure voltage level as low as one microvolt and current as low as one picoampere, he found that all of these different materials exhibited similar noise characteristics at a given temperature. The spectrum of this noise was found to be flat over the full range of frequencies of measurement and was called *white noise*. Over the extent of these measurements, the resulting spectrum was similar to the shape of the spectrum of the unit impulse of Figure 10.3. Johnson also found that the level of the noise is a direct function of temperature. He was able to construct a mathematical model of the noise that showed that the noise level is

$$N(f,t) = kTB, \quad (10.2)$$

where T is the absolute temperature of the resistor and B is the frequency bandwidth of the measurement. With accurate measurements, he found that the constant k is equal to Boltzmann's constant. This is a rather curious coincidence, since Boltzmann's analytical efforts were in the area of gases and thermodynamics. However, recall that Planck had said that the actions of electrons are very similar to that of gasses. Since the measured bandwidth was only limited by the characteristics of the measurement, it was assumed that the spectral bandwidth is unlimited, appearing as shown below.

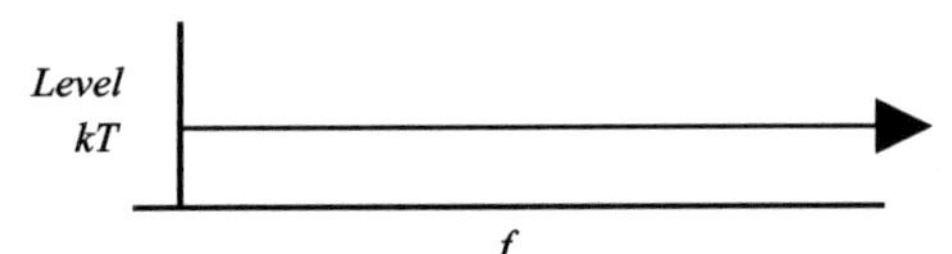

Figure 10.4. Frequency Spectrum of White Noise

Using instruments capable of measuring a wider range of frequencies, Harry Nyquist found that the level of the energy spectrum of resistors noise falls off at higher frequencies (in the microwave range at room temperature). He formulated a mathematical model by adding a "Planck factor",

$$E^2(f,t)\Delta f = \frac{4Rh\Delta f}{\left(e^{\frac{hf}{kT}} - 1\right)}. \quad (10.3)$$

In this equation, E is the voltage across the resistor, and R is its resistance. This equation is in accordance with Johnson's white noise model at the lower frequencies.

Upon re-arranging terms in Nyquist's equation, the result is

$$\frac{E^2(f,T)\Delta f}{4R} = \frac{h\Delta f}{\left(e^{\frac{hf}{kT}} - 1\right)}, \quad (10.4)$$

which can be compared to Planck's state energy equation

$$U(T) = \frac{hf}{\left(e^{\frac{hf}{kT}} - 1\right)}. \quad (10.5)$$

The similarity between these two models indicates that there is a direct relationship between the noise power of a resistor and Planck's state energy function.

The above relationships and the results of extensive measurements provides evidence that electrons behave in a manner similar to that of a gas and that they aıe in continual motion at all but extremely low temperatures. The measured noise of electrical circuits is in the form of random impulses whose shapes vary as a function of the spectral bandwidth of the measuring device or the limitations of the system being measured. The correlation between equation (10.4) and (10.5) provides evidence that Planck's energy state equation represents this type of noise. Therefore, this [model] equation may provide information about the instantaneous actions of electrons in conductors.

Graphical techniques are often used to provide a degree of visualization that allows a greater perception of the relationships between mathematical models and the nature of physical systems. Using this method, the frequency spectrum of Planck's state equation is plotted at three temperatures.

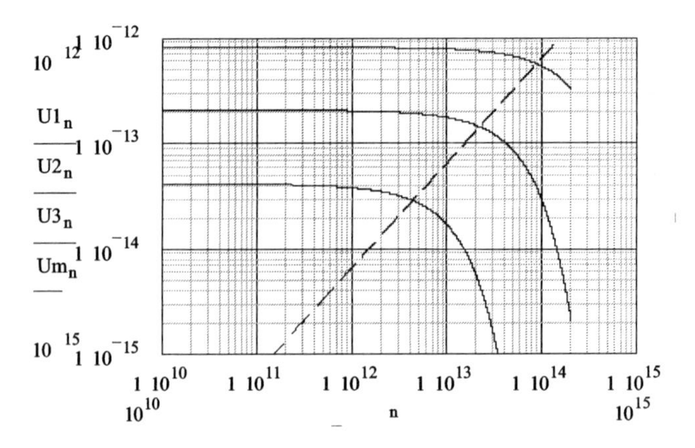

Figure 10. The Frequency Spectrum of Planck's Energy State Equation at 300, 1500 and 6000 deg Kelvin

The dashed line is a plot of $h\nu$ (ν = frequency)

The flat portion of the curves is at the level kT. The dashed straight line curve is a plot of $h\nu$, which intersects kT at the point where $h\nu = kT$. At this point, the level falls off rapidly, and is therefore close to the noise bandwidth. Thus the approximate energy contained in the state energy spectrum is

$$\int_0^{\infty} U(f)df \cong kTf_{int}, \quad (10.6)$$

and

$$f_{int} = \frac{kT}{h} \quad (10.7)$$

is the frequency at which the intersection of the two straight line curves, kT and hf, occurs (note that Planck used both ν and f to represent frequency, and f is used here). The term on the left of equation (10.6) is the total energy, while the term on the right determines the point on the frequency curve of Planck's equation at which the level is

$$U(f_{int}) = \frac{kT}{(e-1)} \cong \frac{kT}{2}. \quad (10.8)$$

This equation shows that the frequency at which the two straight lines intersect is very close to the "half-power point" of the signal spectrum. The total energy is also very close to that of equation (10.6).

Substituting equation (10.7) into equation (10.6),

$$\int_0^{\infty} U(f)df \cong \frac{k^2T^2}{h}. \quad (10.9)$$

The term on the right is a close approximation to the total state energy of the system and the total energy of the impulse function.

Equation (10.9) represents the noise energy dissipated in a resistor. Is it therefore possible that conducting electrons of the atom obey the same electrical laws that apply to resistors? Expressing this argument mathematically,

$$\frac{E^2}{R} \stackrel{?}{=} \frac{k^2T^2}{h}. \quad (10.10)$$

Johnson's noise measurements and Nyquist's equation shows that the noise level is kT over the flat portion of the spectrum, and $E = kT$ in the above equation. Equating the terms,

$$h = \overline{R_e} \quad (10.11)$$

The implication of this equation is that Planck's constant represents the amount of resistance to the movement of the electrons in a conductor. This would imply that "the meaning of h" is the average resistance of an atom against the loss of an electron by the average noise voltage of a conductor. However, h is a very low number as compared to ordinary electrical resistors, and the value of resistance in Johnson's experiment varied over a wide range. At the molecular level, h would necessarily have to correlate with the resistivity of the material, in which case the resistance of a conductor is equal to the length of the conductor divided by its area. There are many atoms along the length of a resistor, and these resistances add, thus accounting for the comparatively high resistance values normally encountered in electronics. Therefore, the resistivity ρ of a material is the important parameter, which varies over a wide range. The equivalent resistivity of h is 0.6 at 320 deg K, while the lowest resistivity of a common metal (silver) at the same temperature is 1.7. The differences in resistivity are much greater for other materials, so there must be another factor involved if there is a correlation between these two phenomena. It is possible that the *internal* forces that produce electron noise act somewhat differently than for an *external* voltage applied to a conductor. The noise voltage is produced by heat in the form of radiation, while resistance is measured by applying a direct current to the conductor. Another possibility is that the probability distributions of the noise voltage varies with the type of conducting materials (Gaussian distributions do not all have the same exact shape). The latter possibility is testable and may also be related to the first possibility. There is also the question as to how the phenomena of superconductivity, which occurs at very low temperatures, relates to these factors. This is an interesting subject that deserves further investigation in subsequent efforts.

The correlation between the characteristics of noise energy and state energy has thus been demonstrated by virtue of the characteristics of their frequency spectra. White (Johnson) noise and Nyquist noise both occur in the form of impulses, as verified by measurements. The Nyquist equation correlates accurate with the spectrum of noise over a wide range of frequencies, and it was shown to be equivalent to Planck's state energy equation. The state energy equation is mathematically integrable and does not contain any undesired mathematical singularities. Therefore, the essential elements are present to derive the shape of the impulse function. The inverse transform of Planck's state energy equation can be utilized to derive the resulting time function. This process is rather complex, so a more simple approximation method will next be used to illustrate the general shape of a noise impulse.

The following material is based on the efforts of an excellent book on the subject: *The Fourier Integral and Its Applications* by A. Papoulis. The first step is to show that random impulses have flat spectra, and then the time function is derived from the inverse transform of the power spectrum. Obtaining the voltage impulse from the power spectrum involves complex and difficult mathematical techniques. The inverse transform of the power spectrum produces a time function that is the *autocorrelation* of the voltage impulse function. Since Planck's state equation does not appear in the tables of transforms, the time function must be derived. The easiest and most straightforward method of deriving the transform of Planck's energy equation is to use numerical analysis, as performed by a software program (Mathcad). The inverse transform of *U*(*f*), thus obtained, is pictured the graph below for three temperatures.

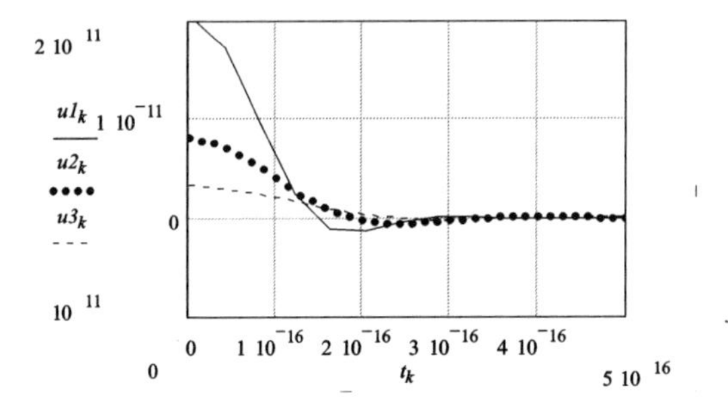

Figure 10.11. Inverse Transform of the State Space Equation At Three Temperatures

The peak level of the energy pulse in this graph increases with temperature, which correlates with the variations in the average power measurements of white noise. The function *u*(*t*) in the above plot represents the autocorrelation of the impulse function $h(t)$ (see Figure 10.1) . A time response function is always narrower than its autocorrelation function (an exception is the unit impulse function, which has no width). Unfortunately, deriving $h(t)$ from its autocorrelation function is not a simple process. The methods of Section 12-4 of Papoulis' book will be utilized.

Certain types of signals produce a flat spectrum

$$S(\omega) = 1. \qquad (10.12)$$

The unit impulse $\delta(t)$ has a flat spectrum, but for impulses that occur at synchronously (at a specific harmonic rate), the spectrum is no longer continuous. Spikes appear in the spectrum at multiples of the repetition frequency. However, when the impulses are *uncorrelated*, the spectrum is flat and continuous, as is the spectrum of white noise. A sequence of numbers α_m are uncorrelated when the following equations hold,

$$\lim_{n\to\infty} \frac{1}{2n} \sum_{m=-n}^{n} \alpha_m{}^2 = 1 \qquad (10.13)$$

$$\lim_{n\to\infty} \frac{1}{2n} \sum_{m=-n}^{n} \alpha_m \alpha_{k+m} = 0. \qquad (10.14)$$

Using the uncorrelated numbers α_n as described by these equations, the following equation represents a sequence of uncorrelated pulses,

$$w(t) = \sum_{k=-\infty}^{\infty} \alpha_k \delta(t-k). \quad (10.15)$$

In this equation, k is the time delay of the pulses, and α_k numbers are randomly distributed and are either positive or negative (a set of such numbers were derived by Wiener). The autocorrelation of a sequence of uncorrelated impulses is

$$R_w(t) = \delta(t), \quad (10.16)$$

and its power spectrum $S(\omega)$ is flat (Papoulis).

$$S(\omega) = \left|H(\omega)^2\right|. \quad (10.17)$$

The next task is to determine the characteristic function $H(\omega)$ and then find its inverse transform

$$h(t) \leftrightarrow H(\omega). \quad (10.18)$$

The response function $H(\omega)$ is not generally uniquely determined from (10.17), since it is a function of the square root of the power function $S(\omega)$,

$$H(\omega) = \sqrt{S(\omega)}\, e^{j\varphi(\omega)}. \quad (10.19)$$

The coefficient $\varphi(\omega)$ of the phasor is arbitrary, and $S(\omega)$ may have complex roots that are not easily obtained. The inverse transform of $H(\omega)$ is unique when $h(t)$ is *causal* (equals zero for negative t). The integral equation of Paley/Wiener can be applied to the power function $S(\omega)$ in order to make this determination, and if it holds $H(\omega)$ will have a causal inverse. However, we already knew that the impulse function is causal from Planck's analysis and the measurements of noise. In the following approach, the response of a single average noise pulse is derived, thus avoiding the considerable complexity of analyzing ensembles of random pulses.

The amplitude of white/Johnson/Nyquist noise pulses varies randomly and has a *Gaussian* probability distribution

$$G(x) = \frac{1}{\sqrt{2\pi}} e^{-x^2/2}$$

Therefore, the amplitudes of the pulses of Figure 10.11 are the average values of the Gaussian distribution.

The occurrence of white noise pulses varies randomly, and the probability of the presence of a pulse over a small time interval t_a within a large time interval T is described by the *Poisson* distribution

$$P\{k \text{ in } t_a\} = \binom{n}{k} p^k q^{n-k} \simeq e^{-nt_a/t} \frac{(nt_a/T)^k}{k!} \simeq e^{-nt_a/T} \frac{(nt_a/T)^k}{k!}. \quad (10.21)$$

This probability distribution applies to a random *noise process* (note the similarity to Planck's probability equation from which the energy state equation was derived, and also the differences). In determining the inverse of the power function, the point of occurrence of the noise pulse in time is chosen to be the beginning of an impulse, which is a simplification that permits the analysis of a typical pulse over a comparatively small time interval *T*. Papoulis used this method, which avoids the difficulties associated with manipulating the probability distributions in the mathematical process. A random process simply produces a solid spectrum, and equation (10.21) can be used to determine the average pulse rate.

The autocorrelation pulse of Figure 10.11 represents an average noise power pulse at a given temperature, and it has been shown that the random occurrence of these pulses produces a solid spectrum that correlates with Planck's energy state equation. The next step is to derive the voltage pulse from the power pulse. Some of the general characteristics of autocorrelation functions will provide insight as to the expected characteristics of noise voltage impulses. The autocorrelation of a function with finite energy is defined by an integral equation.

$$\rho_h(t) = \int_{-\infty}^{\infty} h(t+\tau)h(\tau)d\tau. \quad (10.22)$$

The above autocorrelation function $\rho_h(t) = u(t)$ is the inverse transform of Planck's energy state equation. Using the terminology that applies to Figure 10.1.

Since $h(t)$ is unknown, it must be derived from its autocorrelation. It is possible that the shape of the impulses may not be unique, as was discussed above. Graphical methods can be used to derive an approximate solution or to verify a result. As an example, the autocorrelation of the unit impulse of Figure 10.2 is illustrated below,

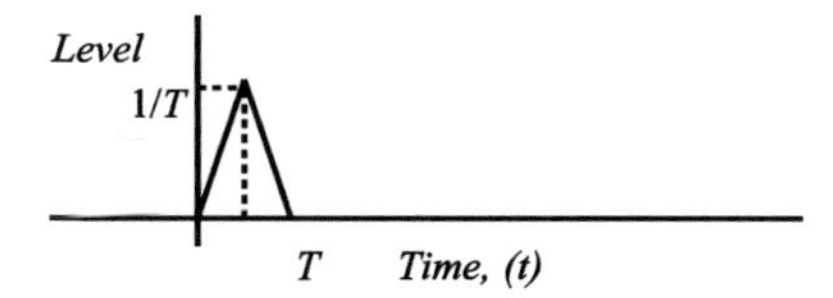

Figure 10.12 Autocorrelated Unit Pulse Function vs Time

The autocorrelation of a causal rectangular pulse produces a triangular pulse that is widened in time. The Fourier transform of the rectangular pulse is

$$F(\omega) = \frac{2\sin\omega T}{\omega}, \quad (10.23)$$

and its energy spectrum is

$$U(\omega) = |F(\omega)|^2 = A^2 = \frac{4\sin^2\omega T}{\omega^2}. \quad (10.24)$$

The square root of the above equation is easily obtained, but there are two roots, one positive and one negative. The phase term that appears in equation (10.19) allows the choice of either root as a solution. In the method of Papoulis, the positive root conforms to the real solution.

The energy state equation has a transcendental function in the denominator, and solving this equation will not be attempted at this time due to the mathematical difficulties associated with these functions. Instead, an approximate solution is shown in Figure 10.13. This function was derived from the response of a typical multiple pole electrical network, and the parameters were varied until the time function fitted the expected time response of an autocorrelation function.

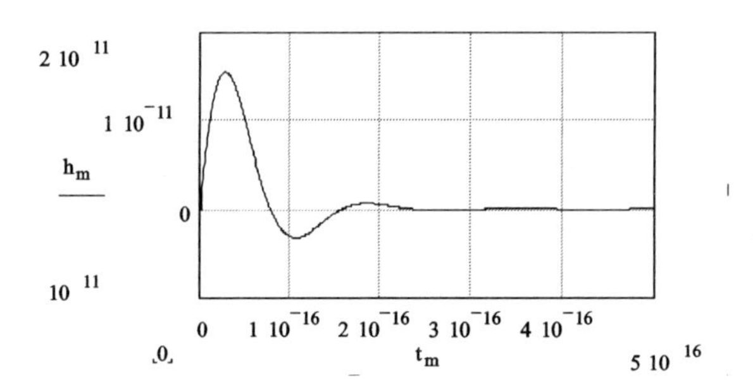

Figure 10.13. Approximate Shape of th› Random Time Pulses of Noise As Derived from Planck's Energy State Equation ($T = 6000$ deg K)

Note that the above pulse is narrower than the autocorrelation pulse, which is a typical result for the autocorrelation process, and it has a shape that would be expected for the inverse of the autocorrelation process. The pulse width is extremely narrow as compared to signal pulses that have been yet synthesized in the laboratory and verified by direct measurements, which is the desired result. Therefore, the above (realizable) pulse signal can replace the mathematical impulse function of Heaviside for all practical applications, thus avoiding the mathematical difficulties associated with singularities. The amplitude, pulse width and power are finite, unlike the unit impulse function. There is an abrupt change in level of the above pulse that occurs at the origin, which is a slight error in the model that represents a comparatively small amount of noise pulse power. Eliminating this slight error is not so easy, however, since the analytical process involves nonlinear and time variant functions, so that will be left for another day.

The general shape of a random noise pulse, as determined from the inverse transform of the energy state equation, has been illustrated. According to Papoulis, a series of time pulses produce a solid spectrum, which is the case for both Planck's state equation and the characteristics of white noise. Planck's assertion that electrons behave as a gas is supported by the above analysis[2]. There are many ways to provide additional verification of these results that will not be covered here. To go further in deriving a more accurate pulse shape involves mathematical analysis that becomes more complex. For instance, the square root of the energy equation is an irrational function, and the inverse Laplace transforms of a similar class of irrational functions are Bessel functions. These Bessel functions are solutions

2 T.L. Martin, *Physical Basis for Electrical Engineering,* p. 168, The "perfect gas model" pictures electrons behaving as a gas, just as Boltzmann and Planck had asserted.

of a differential equation whose coefficients vary with time. The solutions to differential equations of this type are generally not easily obtained[3], even with highly sophisticated computer programs that employ highly accurate numerical analysis. Therefore, Bessel's equation is probably not the best choice to represent electron noise. If the proper time-variant differential equation can be formulated, it could lead to a mathematical solution more easily obtained in closed form.

The relationships between Planck's theory and Heaviside's methods have thus been presented. The theories correlate with the measurements of electrical noise and the known characteristics of the signals produced by electrical circuits. Heaviside's method falls within the category of those theories that do not change with time, as described by Planck.

In the next section, the properties of electrical fields and radiation are examined.

---

3 *The Fourier Integral and Its Applications*, A. Papoulis, Chapter 12

Planck's Columbia Lectures

Electromagnetic Radiation

It is rather surprising that the true nature of radiation has never been fully determined. In particular, various measurements show that radiation is a transverse phenomenon, while other measurements indicate that electric fields propagate radially. Maxwell's equations have been interpreted on the basis of a radial wavefront, while both electromagnetic and light radiation exhibit transverse characteristics. In 2003, I presented a technical paper at the *Antennas and Propagation Society International Symposium* (a division of the *Institute of Electrical and Electronic Engineers*), based on information in my book *Secrets of the Atom*, that provides an explanation for this anomaly.

There have been various theories as to the nature of electrical radiation. Early theories were based on the presence of an *ether* in the universe through which the energy propagates. The physicist Fitzgerald (see Chapter 8) developed an ether theory in which the size of objects vary with velocity. Lorentz's equations were based on Fitzgerald's equation, and Einstein's theory of relativity was based on Lorentz's equations. It is a paradigm that Einstein's theory eliminated the need for the presence of an ether to justify the changes in parameters that vary as a function of velocity (see chapter 8), while utilizing equations that were based on the presence of an ether. Although the ether theory was never disproved, most scientists no longer believe in it.

James Clerk Maxwell developed a set of differential equations that described various characteristics of electromagnetic fields (which Heaviside simplified). Although never proved, these equations work and have been accepted by most scientists. The application of these equations, with restrictions imposed by several assumptions, support the concept of radial propagation of electrical fields. Maxwell was able to derive the exact velocity of propagation c by calculation. The radiating wave is presumed to have an electric and a magnetic component that act at right angles to each other throughout space, and there is evidence to support this conclusion. Imagined pictures of the two components of a radiating wave can be found in various texts. However, none of these illustrations, some of which depict plane waves in the far field (an approximation that applies only over a small area in space), fit the picture of a transverse wave.

There is another model of radiation that is not theoretical. The Mesny equations are based on measurements of the electromagnetic radiation of an antenna. These equations can be found in most authoritative texts on radiation, but they may have first been published in 1936. The electric and magnetic components of the field strength of a radiating dipole antenna are described by the Mesny equations:

$$E_r = -\frac{30I\lambda l}{\pi}\cdot\frac{\cos\theta}{r^3}\cdot[\cos(\omega t - \alpha r) - \alpha r\sin(\omega t - \alpha r)] \quad (10.25)$$

$$E_t = \frac{30I\lambda l}{2\pi}\cdot\frac{\sin\theta}{r^3}\cdot[\cos(\omega t - \alpha r) - \alpha r\sin(\omega t - \alpha r) - \alpha^2 r^2\cos(\omega t - \alpha r)] \quad (10.26)$$

$$H = \frac{Il}{4\pi}\cdot\frac{\sin\theta}{r^2}\cdot[\sin(\omega t - \alpha r) - \alpha r\cos(\omega t - \alpha r). \quad (10.27)$$

Two dipoles are used for the measurements. When the transmitting dipole is oriented vertically, the receiving antenna can then be oriented at various angles to the antenna for the measurements. The first equation represents the radial electric field E_r (receiving antenna oriented along a radial), and the second equation represents the tangential electric field E_t (receiving antenna oriented transverse to the radial). The third equation represents the magnetic field component, which is orthogonal to the tangential and radial field components.

The parameters of these field equations are

r = distance between the antennas

θ = angle between the radial and the vertical

I = current in the dipole

l = length of dipole

λ = wavelength

$\omega = 2\pi f$ = radian frequency

f = frequency

$\alpha = 2\pi/\lambda$.

The length of the dipole is shorter than one-tenth wavelength, which allows calculations of the field strength of large antennas or many antennas.

The above equations show that the electric and magnetic fields of a transmitting antenna vary with time and with distance. Therefore, the field that surrounds the antenna forms a three-dimensional wave that moves with time. A simplification results by orienting the radial angle θ

between the two antennas at 90 degrees. The field strength in the plane orthogonal to the transmitting antenna and passing through its center can then be determined and the field wave plotted. A further simplification is obtained by setting the antenna current to unity. Considering only the far field, the radial electric field level is negligible. The field equations then reduce to the magnetic field and the tangential electric field,

$$E_t = \frac{30\lambda l}{2\pi}\cdot\frac{1}{r^3}\cdot[\cos(\omega t - \alpha r) - \alpha r\sin(\omega t - \alpha r) - \alpha^2 r^2 \cos(\omega t - \alpha r)] \quad (10.28)$$

$$H = \frac{l}{4\pi r^2}\cdot[\sin(\omega t - \alpha r) - \alpha r\cos(\omega t - \alpha r). \quad (10.29)$$

The parameters of these equations are next re-arranged for greater clarity. Substituting the value for α,

$$E_t = \frac{30\lambda l}{2\pi}\cdot\frac{1}{r^3}\cdot[\cos(\omega t - \frac{2\pi r}{\lambda}) - \frac{2\pi r}{\lambda}\sin(\omega t - \frac{2\pi r}{\lambda}) - \frac{4\pi^2 r^2}{\lambda^2}\cos(\omega t - \frac{2\pi r}{\lambda})] \quad (10.30)$$

$$H = \frac{l}{4\pi r^2}\cdot[\sin(\omega t - \frac{2\pi r}{\lambda}) - \frac{2\pi r}{\lambda}\cos(\omega t - \frac{2\pi r}{\lambda})]. \quad (10.31)$$

The radiation equations apply to free space, in which case propagation occurs at the speed of light, and therefore

$$\lambda = \frac{c}{f}. \quad (10.32)$$

Substituting into the above equations,

$$E_t = \frac{30cl}{\omega} \cdot \frac{1}{r^3} \cdot [\cos\omega(t - \frac{r}{c}) - \frac{\omega r}{c}\sin\omega(t - \frac{r}{c}) - \frac{\omega^2 r^2}{c^2}\cos\omega(t - \frac{r}{c})] \quad (10.33)$$

$$H = \frac{l}{4\pi r^2} \cdot [\sin\omega(t - \frac{\omega r}{c}) - \frac{\omega r}{c}\cos\omega(t - \frac{2\pi r}{\lambda})]. \quad (10.34)$$

In the far field $r >> \lambda$, the equations reduce to

$$E_t = \frac{-30\omega l}{c} \cdot \frac{1}{r} \cdot \cos\omega(t - \frac{r}{c})] \quad (10.35)$$

$$H = \frac{-\omega l}{4\pi c} \cdot \frac{1}{r} \cdot \cos\omega(t - \frac{r}{c})]. \quad (10.36)$$

Several observations can be made about the above equations, which are quite similar. The first term of equation (10.33) predominates for $r < \lambda$, so the signal level varies with $1/r^3$ in the near field, while it varies with $1/r$ in the far field. The question is how this could be possible, since space has three dimensions. The level in the near field varies as a function of volume, which relates direction to the three dimensions of space. However, in the far field it varies with only one spatial dimension, $1/r$. So if a one-dimensional field wave exists, then what is its physical shape? Certainly, a spherical wave that expands in a radial direction is defined in terms of three dimensions, which is something more than one would expect from an approximation.

There are several possibilities that could account for

this dilemma. First, the signal level could be a function of the velocity of the wave. However, the radial velocity of propagation is constant in free space. Since the energy fills three-dimensional space, a $1/r^3$ level variation would then be expected, which is also a contradiction since this is not the case.

A second possibility, which is the accepted supposition, is that these waves somehow propagate radially through space and somehow vary with $1/r^3$ in the near field and $1/r$ in the far field. The electric field is transverse to the orientation of the antenna, and the magnetic field is orthogonal to the electric field. In some physics texts, it is argued that one wave alternately creates the other in the far field. However, equations (10.35) and (10.36) are coincident with respect to time and space, which contradicts this argument. The Mesny equations shows that the energy distribution of the field changes dramatically near the edges of a half-wave antenna, changing smoothly from inverse third order to inverse first order variation. Therefore, there is no indication of a breakaway field, such as one might expect for this argument.

This is an indication that the *shape* of the field wave may be contracting, which the third possibility. A fourth possibility is that the field wave varies in more than one direction and one of these velocities increases with radius.

Observations of equations (10.35) and (10.36) and other measurements lead to conclusions that favor the second and/or third possibilities. Near field measurements show that the electric field wave moves back and forth along the antenna as a transverse wave in the near field. The center of this near field wave lies along a vertical line in space that moves back and forth along the antenna. If this wave did not change its shape with radial distance, then propagation of energy throughout space would be instantaneous! This is obviously not the case, so some effect must be occurring that is a function of radius, which begins near the edges of the antenna. Further observations indicate that the electric wave moves back and forth tangentially in the far field.

To illustrate this spatial wave effect, the field waves will be plotted from the far field equation. Equations (10.35) and (10.36) are identical, except for the magnitude of the received signal, and only the electric wave will be examined.

For simplicity, the points lying along centers of the waves will be plotted. The cosine term of equation (10.35) is the electric wave equation

$$e(t) = \cos\omega(t - \frac{r}{c})]. \quad (10.37)$$

A time plot of this equation is illustrated below:

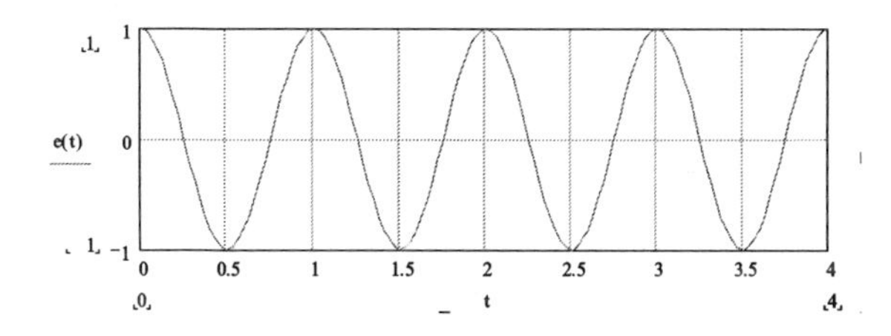

Figure 10.14. Time plot of a Cosine Wave

The frequency is normalized ($f = 1$), and the radius r is near zero (close to the antenna) in this plot. The center of the wave occurs at points where e(t) = 0, and the curve repeats itself periodically at 0.75, 1.75, 2.75, etc. . As the radius is increased, the curve moves to the right , with a corresponding time delay equal to

$$t_d = \frac{r}{c}. \quad (10.38)$$

The derivative of this equation dr/dt is the velocity of propagation c, the speed of light.

The cosine wave of magnetic wave equation (10.37) also varies with time. If this was a radial wave, rather than a transverse wave, then the measurements would support the concept of spherical wavefront. However, the radial component is negligible in the far field, which is another contradiction. Clearly, the transverse waves must be moving back and forth along a circular path (tangentially) at any given radius. This is most easily seen from the eigenvector equation

$$A \cdot e(t) = A \cdot \cos\omega(t - \frac{r}{c})] = A \cdot \frac{(e^{j\omega(t-\frac{r}{c})} - e^{-j\omega(t-\frac{r}{c})})}{2}. \quad (10.39)$$

There are two eigenvectors in this equation, and A is the amplitude of the vector, while +/- jω(t - r/c) is the angle of the “ray” and r is its length in space. A plot of the first eigenvector is plotted below:

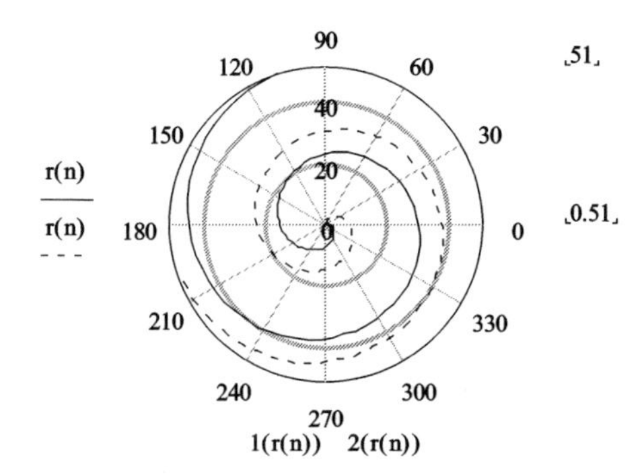

Figure 10.15. Plot of the First Eigenvector of Equation (10.37) At Two Points in Time

The solid line in the above graph is a plot of the wave at one point in time, while the dashed line represents the wave at a later point in time. Therefore, the first eigenvector is rotating counterclockwise. The exponent of the second eigenvector is negative, and it rotates in the opposite direction. When a third vector that is delayed by the same increment in time is plotted, the radial spacing between any two curves is identical, and the radial difference divided by the time difference is equal to the speed of light c. The wave is moving in two directions at the same time, one in the radial direction and the other in the tangential direction. The tangential velocity of the waves is equal to

$$v_t = \omega r. \quad (10.40)$$

Thus, the tangential velocity increases with radius, while the radial velocity of electromagnetic waves is fixed and equals the speed of light c. As it turns out, the tangential velocity of the radiating wave equals the speed of light just beyond the edges of a half-wave dipole and increases indefinitely with the distance from the antenna!

A plot of the centers of two eigenvectors of equation (10.39) at two points in time is depicted below:

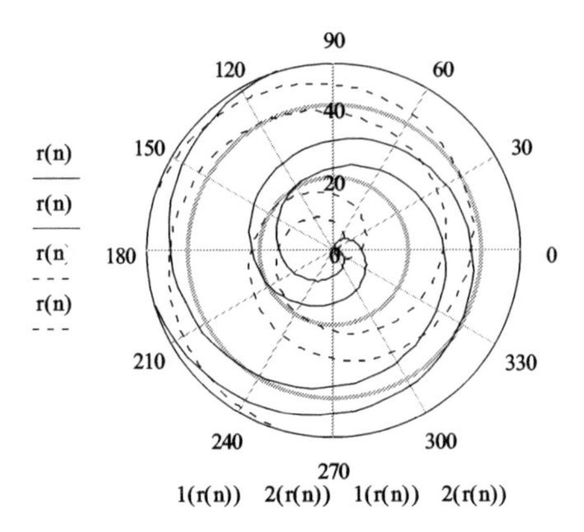

Figure 10.16. Plot of the Two Eigenvectors of Equation (10.37) At Two Points in Time

At the points where the two eigenvectors (represented by the solid and dashed lines) cross, they cancel since they are of opposing polarity. Conversely, where they are in opposition they add together to form a maximum. This accounts for the broad lobe that is typical for a dipole antenna. Greater complexity results when the vectors are plotted at many points in time, which can be observed in the technical paper referenced above.

Evidence has been presented that indicates that the electric field waves bend with tangential velocity, and the amount of bending becomes significant in the far field. The bending of the wave is expressed by equation 10.38, which represents the time delay and a rotational phase delay. The delay is a linear function of radius, which correlates with the fixed speed of light.

The fact that the tangential electric force wave bends means that there must be come opposing force that tends to resist the moving wave. This phenomenon is what would be expected in the presence of a medium, such as an ether, that permeates the universe. Lord Kelvin pictured a similar result in his description of the "luminiferous ether". He claimed that the electromagnetic waves are much stronger than steel and much more flexible. In demonstrating the effect, he waved a stick back and forth in a bowl of jelly and created waves that he said are similar to radiation. The curves of the waves that he drew are circular and appear very similar to the ones derived from the above antenna wave equations.

If there is an ether present in the universe, then it must have a viscosity. The resulting calculation yields the following value for the viscosity of space:

$$b = \frac{1}{c} = 3.333 \cdot 10^{-9}. \qquad (10.41)$$

This is an extremely low value, even compared with the least dense matter in the universe, hydrogen, which fits Lord Kelvin's description of the luminiferous ether.

Another observation is that the radial electric and magnetic field equations, (10.25) and (10.27), are embedded within the tangential electric field equation (10.26). Furthermore, far field equations (10.26) and (10.27) are similar, except for the magnitude constants. Is it then possible that the magnetic field and the electric field are two properties of one field, the electric field? As observed above, the electric and magnetic field waves are coincident throughout time and space in the far fie.d [see equations (10.35 and 10.36)].

In Planck's example II of Lecture 7, the Principle of Least Action was applied to a system in which the coordinates of state form a continuous manifold. Planck's last example of the Principle of Least Action was for the laws of motion of perfectly elastic bodies with displacement components $\mathfrak{v}_z$, in the special case of electrodynamic processes in a vacuum. He derived Maxwell's equations as based on his equation (64) for the expressions for the electric and magnetic fields. These equations are of the following form:

$$\mathfrak{E} = -a\dot{\mathfrak{v}}, \quad \mathfrak{H} = b\,\mathrm{curl}\,\mathfrak{v}, \qquad (10.40)$$

where a and b are constants that depend on the characteristics of the medium. Planck solved these equations by substitution and found that

$$\dot{\mathfrak{E}} = \sqrt{\frac{\epsilon h}{\mu k}}\,\mathrm{curl}\,\mathfrak{H}. \qquad (10.41)$$

This equation shows that the magnetic field is directly related to the derivative of the electric field. The transverse wave equation contains the terms of the radial electric and magnetic wave equations, whereas neither the radial electric nor magnetic wave equations contain all of the terms within the tangential electric wave equation.

Equation (10.41) provides an indication that the magnetic properties of an electromagnetic wave are a function of the derivative of the electric wave. In the next section, this property will be investigated further.

There is other supporting evidence for the above arguments. The phasors of equation (10.39) represent the electric field, and upon differentiation,

$$\frac{d}{dt}(e^{j\omega(t-r/c)}) = j\omega e^{j\omega(t-r/c)} = j\omega e^{j(\omega t-\varphi(t))}, \qquad (10.42)$$

which shows that the derivative of the electric field is in a direction orthogonal to the radius vector, which fits the definition of curl. The phase delay of the eigenvectors of equation (10.39) is equal to $\omega r/c$, and the direction of the wave is tangential by equation (10.42). Therefore,

$$j\varphi(t) = j\frac{\omega r}{c} = j\frac{v}{c}, \qquad (10.43)$$

which is equivalent to the imaginary angle of the Minkowski transformation that forms the basis of Einstein's general theory of relativity (see Lecture 8). Therefore, equation (10.43) gives meaning to the imaginary term in the Minkowski transformation. Clearly, the velocity of the rotational wave varies indefinitely with radius and exceeds the speed of light in the far field.

The antenna radiation equations, which are based on

measurements, fit together with Maxwell's equations, the Principle of Least Action and the Minkowski transformation. The interpretation of Maxwell's equations can be misleading in picturing the shape of electromagnetics waves in space, which leaves the interpretation open for conjecture. Further, the concept of radial waves in space does not conform with the radiation equations, the transverse electric wave or the Minkowski transformation. This interpretation of the Mesny equations leads to the conclusion that all radiating fields are exclusively electric fields. The graphs of the moving force fields picture waves that move tangentially at speeds far in excess of the speed of light. This phenomenon accounts for the alleged "instantaneous communication" between moving atoms (it is not quite instantaneous).

Planck's Columbia Lectures

A New Model of the Hydrogen Atom:

In my second book, *Secrets of the Atom*, a new model of the hydrogen atom was presented. The hydrogen atom was analyzed as an electrical circuit, since it has just two electric charges: one *electron* and one *proton*. The electron is moving in an orbit around the proton, which constitutes an electronic circuit. Therefore, I had decided to apply the analytical methods of electronic analysis in investigating the properties of the atom. Using the length of path of the electron to calculate the path inductance, and with the attractive force between the two charges to calculate the capacitance, the resonant frequency was determined to be 10^{18} Hz. It is more than coincidence that this is equal to the electron frequency (beta rays exhibit wavelengths that match this frequency). The velocity of the orbiting electron, at this frequency of revolution, turns out to be very close to the speed of light. This velocity is higher than the calculated values found in most physics texts, but there other evidence was found to support this value that is also presented in the book.

In determining the magnetic characteristics of the rotating electron of the hydrogen atom, Maxwell's equations (cross products) and the basic laws of magnetism were utilized. However, there was the problem of defining the magnetic properties and the electric current of a single moving charge. The approach that was used to surmount this difficulty turned out to be similar to Planck's Principle of Least action in Lecture 7. I had not yet studied Planck's lectures prior to making this analysis, and it was a pleasant surprise to discover that Planck had used similar method.

The Principle of Least Action turned out to be a

precursor of electromechanical network analysis. With the additions of the methods of Heaviside, Fourier and Laplace, it becomes a very sophisticated way to analyze complex electronic (and mechanical) circuits. The properties of this method will be summarily reviewed. The fundamental equation concerning the action of a physical process in which an external action produces a movement of the system from one position to another is

$$\int_{t_0}^{t_1} (\delta H + A)\, dt = 0. \quad (10.44)$$

This is equation (57) in Lecture 7, wherein A is the increase in energy that the system experiences in the displacement δ, and H is the kinetic potential. The right hand term A represents the sum of the force x distance components for each of the coordinates. The kinetic potential can be both mechanical and electrical. For a purely electrical system, the kinetic potential is

$$H_\epsilon = \frac{1}{2} L_{11}\dot{\epsilon}_1^{2} + \sum_{n=2}^{N} L_{1n}\dot{\epsilon}_1\dot{\epsilon}_{n} \quad (10.45)$$

where L is the self inductance or mutual inductance, and ϵ is the "cyclical potential", which is the electric charge that is now commonly denoted as q. Therefore,

$$\dot{\epsilon} = \frac{d\epsilon}{dt} = \frac{dq}{dt} = i(t) \qquad (10.46)$$

is the electrical current through the inductance,

$$H_{\epsilon} = \frac{1}{2} L_{11}{}^{2} + \sum_{n=2}^{N} L_{1n} i_1(t) i_n(t)_{\text{rr}} \quad (10.47)$$

In a similar manner, the mechanical force on the first inductor is

$$\frac{\partial H}{\partial \varphi_1} = \frac{1}{2}\frac{\partial L_{11}}{\partial \varphi_1}\dot{\epsilon}_1^{\,2} + \sum_{n=2}^{N} \frac{\partial L_{1n}}{\partial \varphi_1}\dot{\epsilon}_1\dot{\epsilon}_n = \frac{1}{2}\frac{\partial L_{11}}{\partial \varphi_1} i_1(t)^2 + \sum_{n=2}^{N} \frac{\partial L_{1n}}{\partial \varphi_1} i_1(t) i_n(t). \quad (10.47)$$

This process can also be applied to the self inductive electromotive forces and the resistive energy loss of Planck's method. In general, the contemporary analytical methods of electronic analysis, in which charges are in motion, utilize the notion of electric current *i*(*t*), rather than electronic charge *q*(*t*). The current method works quite well when there are many charges in motion, but not very well when a small number of charges are moving. The Principle of Least Action especially applies to energy concepts, and Planck showed that it can be used to analyze the state of a body, using the example of the thermodynamics of a gas.

In the hydrogen atom, the electron has been assumed to be moving through space in an orbit that is assumed to be circular, while the orbitals are assumed to be spherical. The orbitals are of little use in the analysis of a dynamic system, so the circular orbit is used for the model. When a particle moves through a magnetic field, an orthogonal force is exerted on it:

$$\vec{F} = q\left(\vec{v} \times \vec{B}\right), \quad (10.48)$$

where B is the magnetic flux density and the arrows indicate spatial vectors. Adding the radial attractive electrostatic force E_e results in the equation for the Lorentz force,

$$\vec{F'} = q\left[\vec{E'} + (\vec{v} \times \vec{B})\right]. \quad (10.49)$$

For an electromagnetic field, the flux density is proportional to the current in a loop, resulting in

$$B = \frac{\mu_0 I}{2\pi d}, \quad (10.50)$$

where d is the diameter of the loop. The length of the loop is the sum of a large number of small segments, ds. In order to express the current in terms of the charge, the following equation is used:

$$i(t) = \frac{dq}{dt} = \frac{dq}{ds} \cdot \frac{ds}{dt} = \frac{dq}{ds} \cdot v(t), \quad (10.51)$$

where $v(t)$ is the velocity of the electron. These three

equations were used to solve for the force exerted on the electron,

$$\overrightarrow{F'} = \frac{q^2}{4\pi\varepsilon_0}\left\{r_0 - \frac{\vec{r_0}\times(\vec{v}\times\vec{r_0})}{c^{\to 2}}\right\}. \quad (10.52)$$

This vector equation represents the dynamic force model of the hydrogen atom. At equilibrium, there can be no radial velocity, and the equation reduces to

$$F = \frac{q^2}{4\pi r^2}\left(1 - \frac{v^2}{c^2}\right), \qquad (10.53)$$

which defines the radial force on the electron. The details of these calculations are provided in *Secrets of the Atom.* The maximum level of this force is equal to the constant term of equation (10.53), which is the electrostatic force between the two charges. The variable within the parenthesis represents the variation in the force with electron velocity. It is observed that the force on the electron reaches zero as the velocity of the electron reaches the speed of light. Therefore, the speed of the electron reaches the speed of light in a stable atom.

Note that the electric potential between the two charges governs the action of the electron, just as was the case for the transverse radiation wave. Also note the v^2/c^2 term, which also appears within the radiation equation and in many other important equations in physics. Equation (10.52) represents the basic dynamic model of the hydrogen atom, and equation (10.53) its stable state.

The above model of the hydrogen atom provides an explanation as to why the electron is always attracted to the proton but never crashes into it. As the electron is accelerated toward the proton at ever-increasing speed, an orthogonal force is exerted on it that causes it to also move in orthogonal direction, thus forming an arc in space. It moves in a spiral toward the proton until it reaches the speed of light, moving in a circular orbit.

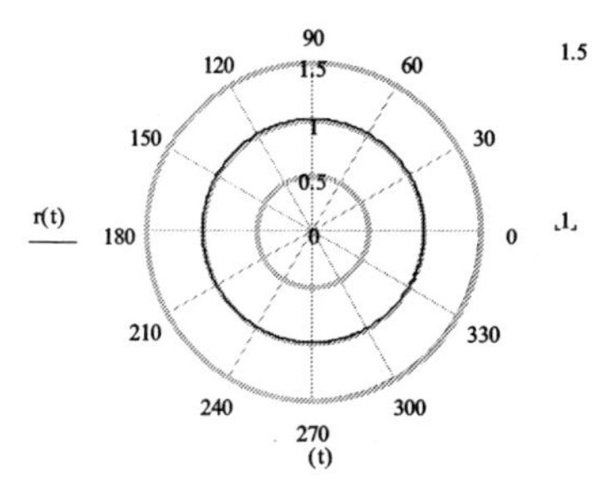

Figure 10.17. Circular Orbit of the Stable Hydrogen Atom

Control system engineers may marvel at the simplicity of this automatic control system, which consists of only two components. The tangential velocity of the electron in a stable orbit is equal to $\omega_0 r_0$, and solving for the frequency using $5 \cdot 10^{-11}$ meter for the radius of the hydrogen atom,

$$f_0 = \frac{c}{2\pi r_0} = 0.9367 \cdot 10^{18}, \qquad (10.54)$$

which is very close to the frequency that was obtained

using other analytical methods in my earlier books.

The results of the above analyses do not result in known contradictions with the real world measurements, and the various methods fit well together. The reason for the earlier estimate of a much lower orbital velocity, as mentioned above, is most likely due to the above forces acting on the electron as it leaves its orbit.

The presence of an ether in the universe also fits in with the concept of the existence of *dark matter*, since the ether must have substance in order to create an opposing force to radiating waves. The above calculation for the viscosity of space is quite low, and although the viscous force is therefore extremely small, which is characteristic of a very low mass density. Also, the amount of "open space" in which the ether is present is much greater than the volume of solid objects, as is observed in telescopic observations of outer space. Therefore, the cumulative mass of the ether may fit that of the mass of dark matter. It will be interesting to see whether or not the above calculated viscosity of space correlates with the estimates for the density of dark matter.

The low viscosity of space indicates that a very high field wave velocity, greater than the speed of light, is required in order to distort the shape of the electric field wave, as pictured in the above wave plots. The analysis shows that the tangential velocity reaches the speed of light at a point very close to the radiating body and increases in proportion to distance. Further, the radiating wave is <u>not</u> spherical in this model, which throws out most of the calculations in currently accepted scientific models of radiation and gravity that depend upon this assumption. Contradictions between any models can be caused by improper assumptions, and the assumption of a spherical radiation wavefront is one that must certainly be questioned.

It is easy to prove that electric fields must bend with motion, for if this was not the case, then instantaneous transmission of information would be possible. This fact is key to my new gravitational theory, which depends upon the electric force equations, which have a higher order falloff as a function of radius in the near field.

I would like to leave you with a few final thoughts. It has been estimated that some 80% of the universe is made of hydrogen. Most of the remainder of the mass is helium, which contains two electrons and two protons, plus two neutrons in its nucleus. Splitting apart a neutron produces an electron and a proton, which indicates that it, too, is electric in nature: perhaps a "compressed hydrogen atom" as described in *Secrets of the Atom.* the coordinate system is allowed to vary. Based on this criteria, all atoms can be viewed as various combinations of hydrogen atoms, which have balanced ionic charges. Hydrogen contains two ions: a proton and an electron. The field around these electric forces controls their actions. These fields move as the charges move, and they change shape as they move. This phenomena is also related to the theory of relativity, which, as Planck stated at the beginning of his last lecture:

"It is my purpose to discuss with you today an hypothesis which represents a magnificent attempt to establish quite generally the dependency of the kinetic potential H upon the velocities, and which is commonly designated as the principle of relativity."

Most of the universe is electric in nature, and the electric charges within atoms, which are in motion, are characterized by their electric fields. In relativity theory, time and space change with motion. This is not the only possibility, as there is another valid choice that I have described in my prior books. The question is the interpretation of the Lorentz equation.The studies of electrical engineering cover the motions of electric charges and their external fields, and the application of these classical methods to the studies of mass and gravitation may soon become a fertile field of study.

This brings us to the end of this chapter, which summarizes some of the results of my efforts over the past two decades. Acceptance of the above analysis did not come easily for me, since I am known as a skeptic. However, the above evidence is very strong. Some theories, such as those of astrophysics have changed radically over the past ten years. In contrast, Planck's radiation theory and Heaviside's method have not changed over time, which is in accordance with Planck's assessment of the most rigorous requirement for a proper scientific theory. It will be interesting to see whether or not the theories, briefly described in this chapter and my earlier books, hold up over time and meet the criteria of Planck's definition of an "invariant".

Planck's Columbia Lectures

Alphabetical Index